Handbook for
SPACE COLONISTS

Other Books by G. Harry Stine

Bits, Bytes, Bauds and Brains
The Silicon Gods
The Hopeful Future
Confrontation in Space
Space Power
The Space Enterprise
The Third Industrial Revolution
Earth Satellites and the Race for Space Superiority

(Science Fiction as Lee Correy)

Manna
Space Doctor
Shuttle Down
Star Driver
The Abode of Life
Contraband Rocket
Rocket Man
Starship Through Space

Handbook
for
SPACE
COLONISTS

G. Harry Stine

with illustrations by
Rick Sternbach

An Owl Book

Holt, Rinehart and Winston / New York

Published by Holt, Rinehart and Winston,
383 Madison Avenue, New York, New York 10017.

Published simultaneously in Canada by Holt, Rinehart and
Winston of Canada, Limited.

Library of Congress Cataloging in Publication Data

Stine G. Harry (George Harry), 1928–
    Handbook for space colonists.

    "An owl book."
    Includes index.
    1. Space flight—Popular works.   2. Space
colonies—Popular works.   I. Sternbach, Rick.
II. Title.
TL793.S758   1984   629.47   84-19830
ISBN 0-03-070741-2 (pbk.)

First Edition

Designer: Gerry Burstein/SOHO Studio

Printed in the United States of America

10 9 8 7 6 5 4 3 2 1

ISBN 0-03-070741-2

EVA (Extravehicular Activity) patch shown on front cover
courtesy of Hamilton Standard, a division of United Technologies

To Pip and Fred Durant

Contents

Handbook for
SPACE COLONISTS

1

Space Is for People

From now until the stars grow cold, human beings like yourself will be living, working, playing, exploring, and traveling in space beyond the Earth's atmosphere.

We're the first species to leave Planet Earth and expand the horizons of our existence into the infinite realm of the universe, because we're the first human generation that's developed the capability to do it. We seem to have been working and building toward this accomplishment throughout human history. Not everything we've done has been a step toward space, but the long-term trends all point toward what Robert A. Heinlein calls the Great Diaspora.

The Purpose of This Handbook

This handbook is intended to help people like yourself when you go into space to work and live. Through the years, you will probably spend a large portion of your life off this planet. It's also a useful handbook for space tourists who plan to spend a few days to a few months in orbit. Like other handbooks and manuals, it's a primer to be studied before you go and a reference book to be referred to from time to time once you're there. Its similarity to existing handbooks and manuals for earthbound living and working ends there because you yourself will be contributing to the changes and new sections that will appear in future editions.

This book will point out what we already know about living and working in space, the type and magnitude of the various hazards and risks involved, and how you can reduce the risks or overcome the hazards. There's no reason why you should "reinvent the wheel" by making the same deadly mistakes as your predecessors who conquered the air and made the first steps into space.

You'll face both physical and psychological problems because the natural environment of space is far different from anything people have encountered on Earth. But you won't live in the space environment because you can't, even though the space environment will be around your spacecraft or habitat at all times. You'll take a small piece of the familiar and comfortable terrestrial environment with you into space, just as we've taken it with us when we've journeyed deep under the Earth's oceans and high into the Earth's atmosphere.

Fortunately, you can call upon the great legacy of experience in such analogous situations here on Earth. Some of the problems of space habitation can be solved by the proper application of known or anticipated technology. But because we don't yet know what all the difficulties will be, you'll still face unique problems for which there are no solutions in these early years of space habitation. You're among those who will discover these new problems because you're pioneering.

Another purpose of this handbook is to help you explain your plans or activities to people who don't understand why you want to live and work in space instead of staying on Earth where there are a lot of good jobs just waiting for qualified people. It will also help you explain to those cautious homebodies who decry "spending all that money in outer space when it could be used to help people here on Earth," and to deal with the many barriers and roadblocks some of them try to put in your way.

Not everyone wants to go into space or to see other people do it. Many people are afraid of it because it's the unknown. They're more comfortable living and working with things as they are. Don't forget that not all of our ancestors left Europe, Africa, or Asia to come to the Americas; some stayed where they were and even tried to stop people from leaving. Even today, most people on Earth never travel more than twenty-five miles from the place where they were born.

No matter when and where pioneers, explorers, settlers, or colonists set out from their homelands, they had to overcome the opposition of those who remained at home. And even in today's world, "enlightened" by the miracles of modern communication and education, people still exist who resemble their ancient counterparts—those who honestly believed that people shouldn't venture out to sea beyond sight of the coastline because they'd sail right off the edge of the world. Most of these sages either try to discourage exploration that threatens the comfortable status quo in which they've got a vested interest, or they attempt to prevent competition by something new that alters what they're already exploiting.

When the Genoese navigator Cristóbal Colón (whose Latinized international name became Christopher Columbus) proposed sailing westward across the Atlantic Ocean to reach the Indies, every intelligent person in Europe knew that the world was spherical. However, in spite of the fact that the Portuguese had preempted the seagoing trade route around Africa to the Orient and had therefore cut Spain out of an important trade area, the royal commission appointed by King Ferdinand and Queen Isabella to review Colón's proposed project in 1490 responded with a list of rationales whose internal contradictions didn't seem to bother the commission in the slightest:

> His plans to sail west to find a shorter route to India are considered impossible because: (1) A voyage to Asia would require three years. (2) The Western Ocean is infinite and perhaps unnavigable. (3) If he reached the Antipodes (the land on the other side of the Earth from Europe), he could not get back. (4) There are no Antipodes because the greater part of the world is covered with water and because St. Augustine says so. (5) So many centuries after the Creation, it is unlikely that anyone could find hitherto unknown lands of any value.

Today, this report seems hilarious to us because of all its inaccuracies and internal contradictions. In spite of such predictions as that of the Spanish Royal Commission, people have gone ahead and done the impossible, impractical, and inconceivable things anyway.

The human race has always been and is still made up of a large number of staid people who till the fields and mind the store, plus a small cadre of dissatisfied, curious, footloose pioneers who aren't satisfied with things as they are and who believe that they can make things better for themselves. Yet the lives of those who stayed home have been altered beyond imagining by the habitation and development of the far frontiers. It will be no different in the years to come.

The restless people in this small minority are different. They're willing to face the new and the unknown even though they may be afraid of it, and they do this for any number of personal reasons. They believe in "progress," an idea disparaged by many people who are of the stay-at-home kind.

In his nineteenth-century essay "Lord, What Is Man?," Samuel Butler wrote, "All progress is based upon a universal innate desire on the part of every organism to live beyond its income."

Pioneers are people who aren't afraid to try to do this. You're willing to bet that the future you make for yourself will be better than the past, and that this betting on the future will permit you to pay your bills when they come due.

Knowing *why* you're doing something isn't an idle academic exercise. It's the basic foundation that supports what you're doing. Although you may already have developed a philosophy of *why* you want to go into space to live and work, let's go over some deeper reasons for doing it both from an individual as well as from a group viewpoint.

Pioneering in Space

People who live and work in space are pioneers and will be for decades to come. You're different from the ordinary run of people, and so were your ancestors who left ancient homelands to venture into the unknown wilderness. In the next fifty years, you'll represent only a tiny fraction of the human race, but your activities will have a profound effect upon the lives of every other human being.

Space pioneering requires that you use every bit of intelligence, know-how, and mental agility you possess. Every possible branch of human knowledge is involved and will be required. This is no different from the pioneering movements and migrations of the past in another way—a way that many starry-eyed space advocates don't really understand but will come face to face with: John Woods Campbell, Jr., once observed that pioneering involves discovering new and more horrible ways to die.

This point of view was summarized by the English poet and playwright Herman James Elroy Flecker (1884–1915), whose poem "The Golden Journey" proclaims:

> We are the pilgrims, master; we shall go
> Always a little farther—it may be
> Across that last, blue mountain rimmed with snow,
> Across that angry, or that shimmering sea.
>
> White on a throne, or lonely in a cave,
> There lives a prophet who will understand
> Why men were born—but surely we are brave
> Who take the Golden Road to Samarkand.

This is a statement of a philosophy or way of thinking about yourself and the rest of the universe.

Why Go into Space?

People go into space and live there because of personal philosophies that they may not fully grasp themselves, and they are probably totally misunderstood by those who remain on the ground by choice.

Great achievements in space have occurred without a long-term operating philosophy, only to fail once initial success was attained. The Apollo manned lunar landing program of the United States of America is an excellent example. The Apollo program was strictly and totally justified on the basis of national political prestige—which, at its foundation, was motivated by a desire for national security. Beyond getting to the Moon, the Apollo program had no clearly defined philosophy and therefore no long-term goals that were understood by the people who paid the bill, the American taxpayers. So that grand project died once it had attained its very limited objective.

Today we know *why* people such as yourself must go to and live on the Moon and in space. You do it not because of national political ideology, but because you'll be using the universe to produce something of value for you and other people. One of the only purposes of the human race that we know of for certain is that we can and therefore should *use* the universe in which we've evolved. If we cannot or should not do this, the universe will let us know in no uncertain terms if and when we exceed the boundaries of *use* so that it becomes *misuse*. And, in fact, in some areas of technology here on Earth, this has already occurred, and we've learned a great deal as a result.

We've used Planet Earth for millennia. We've perhaps misused it locally because of greed born of the philosophy of living in a world of scarcity or because of ignorance of what we were really doing. Some people would even say we've misused the Earth because we've changed parts of it. But the Earth's environment has been drastically altered by living things since life began. Perhaps the purpose of organic life itself is to change the universe, because it appears to be a mechanism for reversing the tendency toward disordering, or "increase in entropy." If entropy is defined as the tendency for all energy in the universe to decay slowly to a common level as a result of the inevitable losses in all systems, thus reducing the universe to an overall state of disorder, then life—and especially "intelligent life" such as we believe ourselves to be—appears to be reversing this trend by being *antientropic* and creating order out of disorder.

You and I probably will never know whether or not this is true. The end of the universe, if there is to be an end, may not take place for another ten billion years or so, depending upon which of the current theories of cosmology one happens to favor. In the meantime, we'll continue to use what we can of the universe around us.

Thus, if we are to use the universe, space is for people and not just

5

machines. We knew this even as we were sending our machines to help us explore the planets, but some people forgot that these machines were only our tools.

The Unmanned Space Explorers

Automated machines—Ranger, Surveyor, Lunar Orbiter, Explorer, Mariner, Viking, Voyager, and others—were indeed magnificent examples of the extension of human senses into the strange, hostile environment of space. But they were commanded by people back on Earth. When something went wrong with the machines, people sitting back on Earth had to puzzle out ways to fix them by remote control. When the machines encountered difficulties for which their computer brains carried no instructions because the situations hadn't been anticipated by their builders, people back on Earth had to figure out, if

The *Vanguard 1* un-
manned Earth satellite
(NASA)

Mission control,
Houston *(NASA)*

possible, how to modify the operations of the machines to handle the new situations.

The machines did provide us with the new information we desired, but they also raised new questions and produced a lot of human frustration.

We sent Ranger and Surveyor spacecraft to the Moon. The pictures they sent back to us on Earth were exciting . . . and frustrating. What was behind that rock? That distant hill? And was the soil that was analyzed by the Surveyor landers really the same as the soil where the Apollo Lunar Module would land? Astronauts themselves finally got the answers and even brought back parts from a Surveyor lander. The lunar astronauts left unmanned scientific measuring instruments called ALSEP on the lunar surface; the automatic machines continued to work for several years after the astronauts departed and finally quit because they needed repairs and there was no human repairman there to do them.

We also sent Viking spacecraft to land on Mars. The photographs they sent back were exciting . . . and frustrating. Again, what was behind that rock? That distant hill? And did the miniaturized, automated biochemical laboratories aboard the two Viking landers *really* detect some sort of organic life? Or were the responses of the automated machinery only the result of a unique chemical accident? There's no way of knowing without either sending a very expensive and technologically advanced robot to Mars or sending a human expedition to find out for certain. The spacecraft and landers finally stopped working and the people on Earth couldn't tell the machines how to fix themselves because the malfunction had severed the communications link.

7

With all of the problems of robot explorers encountering unforeseen difficulties and eventually needing repairs, why did we build unmanned robot spacecraft and send them into space and to the planets in the first place? There are several reasons:

1. Unmanned spacecraft could be sent to the Moon and the planets using available modified military ballistic missiles in the 1960s and 1970s. These missiles didn't have the capability to carry the weight of manned spacecraft.

2. At that time, the launching of unmanned spacecraft with surplus expendable military missiles was less costly than the development of a space transportation system designed and built around the characteristics and limitations of human beings. It took us more than a quarter of a century to stop stuffing people into long-range rocket-powered artillery shells and to begin using even the rudimentary, primitive NASA Space Transportation System, the Shuttle.

3. Unmanned automatic spacecraft eliminated the risks of sending people into the unknown environment of space. Those people who have always resisted change and progress argue that human life is far too precious to be risked in space when we can instead build machine extensions of ourselves to probe the universe. Some of them even go so far as to say that these machines have become a new intelligent species that we, the outmoded humans, are developing as our evolutionary replacements. They state that only this new computerized species should explore the hostile universe beyond our planet.

These people forgot that machines are just tools, not entities in and of themselves.

A human being in space can do all that an unmanned spacecraft can do and all that it can't, which is 99 percent of everything.

The Need for People in Space

People are absolutely necessary in expanding our space frontiers just as they've been necessary on terrestrial frontiers.

We've already looked at the actual report of the Spanish Royal Commission regarding Cristóbal Colón's proposed voyage. But suppose the advocates of unmanned exploration had been there at the time. Their report on the infeasibility of the manned voyage might have concluded with the following recommendations:

However, because the Portuguese now control the seagoing trade with the Indies around Africa and the infidel Turks block the overland trade routes to Asia, we suggest that Señor Colón's hypothesis might have some merit in providing us with an exclusive new trade route of our own. But we caution that it must be tested first. It is far too risky to send sailors on such a hazardous voyage, provided that sailors could be found who were willing to undertake such a dangerous and unknown

The first manned, controlled; heavier-than-air flight with Orville Wright at the controls of the *Flyer* as his brother, Wilbur, runs alongside at Kitty Hawk, North Carolina, December 17, 1903 *(NASA)*

journey. Therefore, the Royal Commission recommends the development of a crewless exploration ship with a sample-return capability. This vessel would sail westward under automatic control. If it encountered land, it would gather a sample of the soil and reverse its course to return to Spain. We estimate at least fifty years will be required to develop and test this ship by means of a series of short voyages to the Azores, and we estimate the cost to be twenty times the value of the royal jewelry pledged to the manned mission by Her Majesty, Isabella of Aragon.

Let's come closer in time and take another fictitious and completely hypothetical look at what might have happened if the unmanned exploration advocates had existed in the year A.D. 1900. Let's suppose that in that year the United States government looked into the matter of flying machines and decided that scientific evidence showed it might be possible to develop such a device, although no one had done so by 1900. An "Ad Hoc Presidential Committee for the Development of Aero Machinery" presented its findings and recommendations in a long report recommending that:

ten million dollars be appropriated to establish a laboratory under Dr. Samuel Pierpont Langley for research on the development of aerodynes, aeroplanes, or other heavier-than-air flying machines. Since many accidents and deaths in experimental aeronautics have shown beyond doubt that flying is a hazardous activity, the Committee recommends that the first phase be the development of a remotely controlled aeroplane that could reasonably be anticipated to make an initial flight by the 1920s. The Committee feels that by 1950 development may have progressed to the point where the first manned aeroplane flight could be made because by that time aeroplanes would have proven themselves safe for human passengers.

These hypothetical examples sound ridiculous because neither scenario matches the way we humans really go about pursuing hazardous activities. They also tell us something about ourselves. Among other things, we're gamblers and risk takers even with our own lives. Somebody's always willing to take a risk for money, principle, pride, or any number of other human motivations. If the Spaniards had followed the advice of their Royal Commission, much less the recommendations of our hypothetical crewless-sailing-ship committee, they would have been left behind in the quest for the New World by other Italians such as Cabot who were hired by the risk-taking English. Americans would have been puttering in their government-funded laboratory when Santos Dumont succeeded in making a manned flight in Paris in 1906.

We're a crisis-oriented risk-taking species. We thrive not on reason and logic but hunches and illogic. That's why we're going into space and why we'll succeed in space.

Space is for people, not machines. The machines of space must be designed around the people of space, and people come only in one model, *Homo sapiens sapiens*, Mark One, Modification Zero.

There are as many reasons for people to live and work in space as there are for people to live and work on Earth. In fact, living and working on Earth has only been a prelude to doing the same things in the rest of the universe.

Astronaut working in space *(NASA)*

Since human beings are therefore the important factors in space travel and living, what do we know about ourselves?

And since the environment of space appears to be hostile to us, what is the environment that we must have around us in order to function properly?

Once these two factors are known, the technical hardware necessary to permit people to live safely and happily in space becomes subject to engineering solutions.

People have been working on the answers for a long time.

The two-man Gemini spacecraft of 1964–1966 *(NASA)*

2 The Legacy of Space Habitation

Why Is This Important?

Why should you spend any time reading about all of the ideas and proposed space stations, habitats, and colonies that have been put forth by people in the past, especially when the habitats in which people are living in space look so different from these ancient concepts?

The point is simply this: We've been thinking for a long time about living off this planet. We've studied the problems and proposed solutions in light of the scientific knowledge and the technical capabilities of the time. We've written about it and lectured on it. We've tried to look at all the problems to find solutions.

And there don't seem to be many insoluble problems, regardless of the contemporary science and engineering. True, the solutions change as technology changes, and new problems rear their heads as scientific investigation and measurement progress.

You may think you're doing something new and different but, upon further study, you'll usually discover that someone else was there ahead of you. The value of knowing and studying previous proposals and concepts doesn't necessarily lie in discovering new approaches and solutions to current problems, but in keeping yourself from "reinventing the square wheel." One of the valuable characteristics of being human is our ability to pass along to others not only our

Dr. Robert H. Goddard
and his liquid propellant
rocket that made its first
flight on March 16, 1926

successful experiments, but also the details of those endeavors that didn't succeed.

An experiment that fails can be even more valuable than one that succeeds. An experiment is nothing more than asking the universe a question, and the universe always gives a complete and honest answer. The answer may not be the one anticipated, however, because you may have asked the wrong question.

There's always a time factor to be taken into account in any technically based activity. A concept or proposal may be impractical at the time it's put forward because of any number of reasons—technical, economic, "it's too new," or "we don't have any background in that area." If you're really smart and bright, you file it away where it can be found later because conditions change, desires change, needs change, economic factors change, and (most importantly) human knowledge of the universe now doubles every five years.

Knowing where you've been can help you learn where you're going. And knowing what's been done before can help prevent you from making the same mistakes, if you pay attention and use that knowledge.

Tsiolkovski's Vision

Mankind will not remain on Earth forever, but in its quest for light and space will at first timidly penetrate beyond the confines of the atmosphere, and later will conquer for itself all the space around the sun. The murky views which some scientists advocate as to the inevitable end of every living thing on Earth should not now be regarded as axiomatic. The finer part of mankind will, in all likelihood, never perish—they will migrate from sun to sun as they go outward. And so there will be no end to life, to intellect, and the perfection of humanity. Its progress is everlasting.

These words were written in another time in another language in another country by a man who richly deserves the accolade of the Father of Space Habitation. He was Konstantin Eduardovich Tsiolkovski, the son of a Polish lumberjack, born in the little village of Izhevskoye, Russia, in the year 1857. An impoverished teacher of science and mathematics who had been deafened by scarlet fever as a youngster, Tsiolkovski was motivated to consider spaceflight by reading a Russian translation of Jules Verne's *De la terre de la lune* (*From the Earth to the Moon*), one of the pioneering classics of science fiction.

Tsiolkovski first began to consider space travel in 1883, privately published his first work on the subject in 1903—the same year the Wright brothers successfully flew at Kitty Hawk—and finally had his ideas and concepts accepted and his work supported by the Soviet government following the October Revolution of 1917.

Tsiolkovski was not merely a dreamer but lent credence to his thinking by

In 1903 the Russian space pioneer Konstantin Eduardovich Tsiolkovski prophesied multistage rockets, space stations, and interplanetary travel.

means of careful mathematical analysis and actual testing where possible. In 1900, he built the first wind tunnel in Russia and tested winged models in it. He designed a space station and wrote about growing plants in it to provide fresh oxygen for the human occupants. Although he may not have been the first person to think about and propose space habitation, he certainly wrote and lectured prolifically on the subject.

Hale's Brick Moon

As Verne wasn't the first to write about human travel to other worlds, Tsiolkovski's thoughts on space habitats were preceded by Edward Everett Hale's story "The Brick Moon," in *The Atlantic Monthly* in 1869. This was the first manned-space-station science-fiction story.

Hale's fictitious young space colonists calculated the cost of their habitat at $214,729—a sure candidate for a budget overrun even more than a century ago and assuming a constant inflation rate of 7 percent. And the figure is certainly open to question because no one can possibly make a budget estimate of that size which is also accurate to six significant numbers!

Of course, the concepts of Tsiolkovski and Hale are now obsolete in terms of the march of technology. However, Hale did conceive of his manned satellite as being coated with bricks in order to withstand the heating of its passage through the atmosphere during launch. It's most certain Hale didn't envision the high-technology bricks that cover the Space Shuttle Orbiter and perform the same function.

Goddard, the "Moon Rocket Man"

Tsiolkovski is often compared to Dr. Robert H. Goddard, America's rocket pioneer. Although motivated to space travel by Verne's fictional story of the trip

Dr. Robert Hutchings Goddard, the American rocket pioneer, whose work on liquid rockets from 1920 to 1945 forms the foundation for space transportation *(NASA)*

Hermann Oberth, the Transylvanian mathematician whose mathematical treatise on space flight sparked the rocket enthusiasts of the 1920s

to the Moon as Tsiolkovski was, Goddard devoted his time and efforts toward the concepts, development, and testing of the space transportation device that would permit space habitation: the liquid propellant rocket.

Oberth, the Practical Visionary

The modern concept of the manned space station is European, and it was the European dreamers who brought engineering and technology to bear upon what had previously been semi-utopian dreaming and science-fiction. In 1923, a Transylvanian mathematician, Hermann Oberth, published in Germany a small book entitled *Die Rakete zu den Planetenräumen* ("The Rocket into Interplanetary Space") which, in spite of its highly mathematical nature, provided the inspiration and motivation for a young Prussian nobleman, Wernher von Braun, to devote his life to the goal of space travel.

In his pioneering book, Oberth wrote of earth-orbiting satellites. He emphasized the practicality and immediacy of a manned satellite in low polar orbit that would be utilized for earth-oriented observations of the weather, land and water resources, and mapping.

Noordung's "Wohnrad"

It was not until 1929, however, that the first book devoted primarily to space stations rather than rockets was published. The author was an Austrian named Potocnic writing under the pseudonym of Hermann Noordung.

17

Dr. Wernher von Braun, whose engineering know-how, political acumen, and charismatic leadership led to the first manned lunar landings in 1969–1972

Noordung's earth-orbiting manned space station consisted of three separate units—the "living wheel" or *Wohnrad*, the power station, and the observatory.

The Wohnrad was a wheel-shaped rotating station 100 feet in diameter, the predecessor of all rotating space structures. The wheel was spun to create artificial gravity at its rim for living purposes. The observatory unit was not described in detail. An electric cable ran from the Wohnrad to the power station, which Noordung described in vague terms as a large paraboloidal mirror with a set of boiler pipes coiled up at the focus. The power station was what we would now call a "closed cycle thermoelectric" solar power satellite of modest size. The Wohnrad itself was designed in quite thorough detail and, in some respects, greatly resembles some existing serious space-base concepts and designs.

Early British Work

Noordung's Wohnrad apparently served as the basis for the next space-station design carried out by Harry Ross and R. A. Smith of the British Interplanetary Society in 1948. The Smith/Ross station was built around a 200-foot parabolic mirror for supplying solar power to the station. The station was designed in great detail to support twenty-four people primarily engaged in both earth observation and astronomy.

The Collier's Space Program

But on March 22, 1952, for the huge sum of fifteen cents, anyone in America could pick up from any newsstand a copy of *Collier's* magazine with its striking

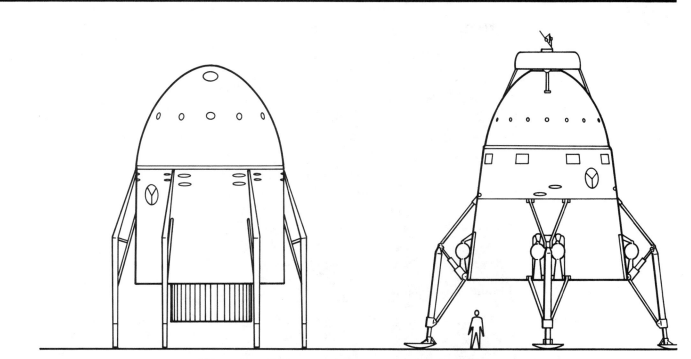

Figure 2-1. Early spaceship designs made by members of the British Interplanetary Society *(Art by Sternbach)*

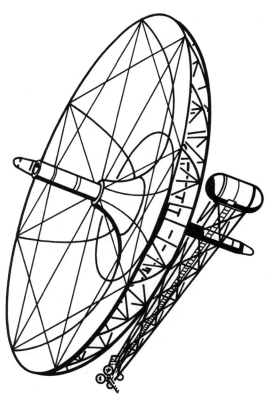

Figure 2-2. One of the first space-station designs carried out by the British Interplanetary Society *(Art by Sternbach)*

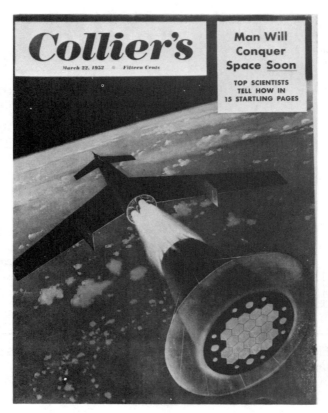

This striking painting by Chesley Bonestell of the staging of a space shuttle was the cover of the 1952 issue of *Collier's* magazine that introduced the series on space travel. (*Collier's*, March 22, 1952)

The three-stage reusable shuttle rocket designed by von Braun for the 1952 *Collier's* series of articles on space travel (*Collier's*, March 22, 1952)

full-color cover painting by the famed space artist Chesley Bonestell, showing the separation of the top stage of a hypothetical winged shuttle spaceship on its way to a polar-orbiting space station, a wheel-shaped doughnut inhabited by eighty people.

In a series of *Collier's* magazine articles and a later series of books in which the individual articles were collected, a group of scientists/engineers/authors led by Dr. Wernher von Braun, Dr. Willy Ley, and editor Cornelius Ryan, with striking and accurate illustrations by the famed space artist Chesley Bonestell, Fred Freeman, and Rolf Kelp, readers could delve into all the details of a working space habitat.

This was the most ambitious attempt ever taken to bring to public attention what the space pioneers had been thinking about for half a century or more. It had a profound influence on space enthusiasts at the time, created a lot of new space advocates among young people, and served as the basis for a series of motion pictures from the Walt Disney Studios.

Hollywood jumped on the space-station bandwagon in the mid-1950s to produce some singularly horrible "bombs" in which the doughnut-shaped von Braun space wheel served as a point of departure for movie art directors, who designed a series of space stations, spaceships, and other space impedimenta that were visually beautiful and technically impossible. But Hollywood has

The rotating space station designed by von Braun and painted by Chesley Bonestell for the 1952 *Collier's* magazine series on space flight (*Collier's*, March 22, 1952)

21

A modification of the von Braun–*Collier's* toroidal space station appeared in the 1954 Paramount motion picture *Conquest of Space*. *(Paramount Pictures Corporation)*

been doing this for decades, is still doing it, and probably will continue to do it because they are, to quote one cinematic impresario, "interested in people, not gadgets."

The Romick Space System

The year 1954 saw the emergence of another manned-space-station concept that exhibited extraordinary technical thoroughness. The work of a group of aerospace engineers working for Goodyear Aircraft Corporation—this company now makes only the famed Goodyear blimps—led by Darrell C. Romick, the proposal took the basic von Braun concept a step beyond.

Romick designed a three-staged earth-to-orbit shuttle rocket with an important feature: each of the three stages was manned with a human crew that flew it back to Earth, thus making the entire shuttle reusable.

Another feature that was borrowed from an earlier concept from the British Interplanetary Society was the proposed use of the fuselage "core" of several shuttle upper stages as the backbone for a unique manned space station. The major portion of the habitat would look, when completed, like a fat sausage 3,000 feet long and 1,000 feet in diameter. On one end of the sausage was a spinning living section, a wheel 50 feet thick and 1,500 feet in diameter. Romick's space station could support a crew of more than a thousand people.

As any good engineer should do, Romick estimated the cost to be similar to

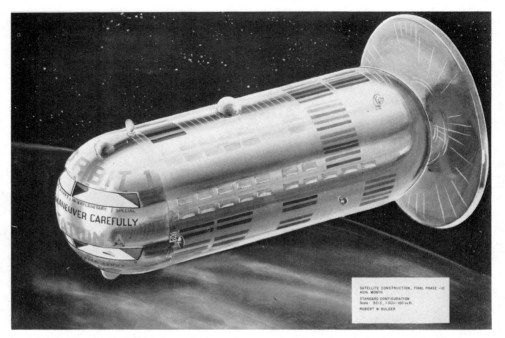

In 1954, Darrell C. Romick and his colleagues at Goodyear designed the first space station made up of modules brought up by winged space shuttles. *(Darrell C. Romick)*

equipping and manning three strategic-bomber wings. His estimated lift costs were $8 per pound of payload to low Earth orbit.

The Romick proposal was unique because it was the first one to combine all the technical and economic factors. But, in common with all space projects, real or projected, it lacked the most important element: the reason to do it at all.

In a great many respects, this 1954 work of Romick and his associates bears a striking resemblance to that done for larger space habitats some fifteen years later. But today few people remember the Romick proposal because, save for a single magazine article in *Mechanix Illustrated* magazine in January 1956, the proposal was primarily seen only by other aerospace engineers and space advocates. It was ahead of its time.

The Apollo Hiatus

The 1950s period of space advocacy triggered by the *Collier's* articles was just as intense on the part of "space cadets" and motion-picture entrepreneurs as that of the 1970s with one exception: in 1952–1955, there was no space travel at all. As a result, space advocates were considered to be wild-eyed fanatics chasing a chimerical dream.

The 1950s era of space advocacy evaporated at the same time that its criticism did: upon the orbiting by the Soviet Union of *Sputnik I* on October 4, 1957. It was followed immediately by a space race with the Soviet Union that

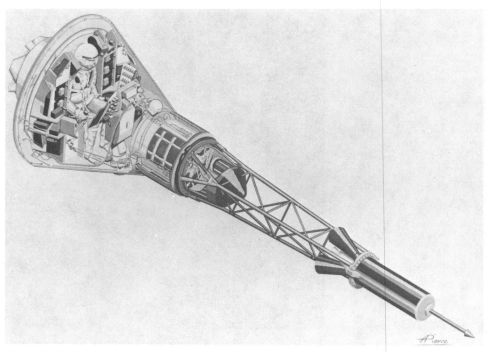

Figure 2-3. The Mercury spacecraft of 1961–1963 *(NASA)*

The three-man Apollo lunar spacecraft of 1968–1972 showing (left to right) the cylindrical service module, the conical command module, and the complex lunar module *(Rockwell International Corporation)*

culminated in 1969 with the *Apollo 11* lunar landing; whereupon the Soviet Union claimed it had never been racing to the Moon at all, and the United States claimed it had won the race.

Although von Braun and his team succeeded in building the huge three-staged manned space shuttle that had been described in great detail in the 1952 issues of *Collier's* (it was called "*Saturn 5*" when it was built), the major space goal of the *Collier's* articles was shifted for political reasons away from the manned space station to the lunar landing. The manned space station was bypassed.

If any space advocate was asked in 1955 to describe the best way to get to the Moon, the answer was pat: Build a space shuttle to get to and from Earth orbit, then build a manned space station in Earth orbit, then go to and from the Moon from the space station. It didn't happen that way. Under the pressure of a time schedule to get a man on the lunar surface in the 1960s, it was decided that the quickest way was to forget the space station and go directly to the Moon, using lunar-orbit rendezvous. This turned out to be a dead end.

Skylab

Although enough hardware had been purchased and built in the United States to support twelve lunar landings, only seven were attempted and six successfully carried out. The leftover Apollo lunar hardware was cleverly adapted to the task of coming back and touching second base, so to speak, by establishing a temporary space station. The last *Saturn 5* moon rocket was used in a two-staged configuration to loft into Earth orbit its third stage, which had been outfitted as a rudimentary manned space station capable of supporting a crew of three people. Three Apollo capsules were used up ferrying three different crews to and from this lash-up known as Skylab.

Carried out on a short budget that permitted no failures, Skylab was a resounding success primarily because of the fact that people could repair the inevitable failures that occurred right from the start.

But Skylab too was a dead end and was abandoned after the third crew visited it. It was allowed to enter the Earth's atmosphere and burn up over Australia while a nonunderstanding news media crowed that the sky was falling.

And the last operable Apollo hardware was used in the 1975 Apollo-Soyuz Test Mission from which the Soviets gained far more in terms of publicity and technical know-how than did the United States. The remaining Apollo hardware was stretched out on the ground to rot in Huntsville, Alabama, Cape Canaveral, Florida, and Johnson Space Center, Texas, because the launch teams had been disbanded and the launch complexes torn down and sold for scrap.

Dandridge M. Cole

On the long-range planning front, however, all was not quiet during the years that Saturn rockets were thundering aloft from Cape Canaveral, Florida, on

The three-man Soviet Soyuz spacecraft. Two of them are shown here docked nose to nose in Moscow's Kosmos Pavilion *(O. Saffek)*

their short-lived orbital and lunar missions. From 1957 to 1965, a quiet and thoughtful dreamer was getting paid for what Tsiolkovski, Oberth, Noordung, and even von Braun had done in their spare time because they were building a dream.

Dandridge MacFarland Cole was a "blue sky" long-range planner and technological forecaster who worked for the Martin Company in Denver, Colorado, until about 1961 when he joined the Missile and Space Division of General Electric Company at Valley Forge, Pennsylvania. Dan Cole was not only a profound and original thinker, but he'd studied what everyone else had done in the way of long-range space planning. His book *Beyond Tomorrow: The Next Fifty Years in Space* was published in 1965 and is a classic today. A collection of his scientific and technical papers rewritten into a popularized style, *Beyond Tomorrow*, has served as a sourcebook for many who followed in his footsteps.

Dan Cole's mind prowled the branching halls of the future. He thought big. He went beyond the simple concepts of space stations and even earth-orbiting space colonies. He saw that there was literally no limit to the size and complexity of space habitats. His concept of "macro life" included not only habitats with ten thousand people, but *mobile* habitats that were huge space vessels with a hundred thousand people living in them. The macro-life ships were huge, self-contained, and independent closed and balanced ecologies.

Dan Cole also investigated the *reasons* why human beings would want to do all this, and came to the same conclusion as the author—or perhaps the conclusion was arrived at during many long discussions that I was fortunate to have with Dan Cole as a close friend and colleague during the last eight years of his too-brief life.

You will be living Dan Cole's great dream doing the things Dan Cole envisioned that people could do in space. He understood the Third Industrial Revolution, space power, space warfare, and Heinlein's Great Diaspora.

In an August 1963 General Electric report, Cole wrote:

The new step in evolution is from man—as the most highly organized example of multicellular life—to the closed-cycle society of macro life. We can already see the beginnings of this new life form in the nuclear submarines and the undersea laboratories, in the space capsules and space simulators, and in the plans for interplanetary vehicles and extraterrestrial colonies. The most highly advanced forms of macro life which may develop in the next fifty years are the the colonies of perhaps 10,000 people in *Queen Mary*-sized space ships or the asteroid colonies. These highly developed societies can cruise indefinitely in the solar system or even venture out into interstellar space. They will constitute new giant life forms by any standard tests now in use in that they will have the capability for motion, growth, reproduction, reaction to stimuli, and even intelligent thought. They will exceed in power and survival value any previous product of evolution known to us here on Earth, and will be the undisputed masters of the known universe. They will be

Figure 2-4. The interior of the "hollow asteroid" space colony visualized by Dan Cole (Martin Marietta)

27

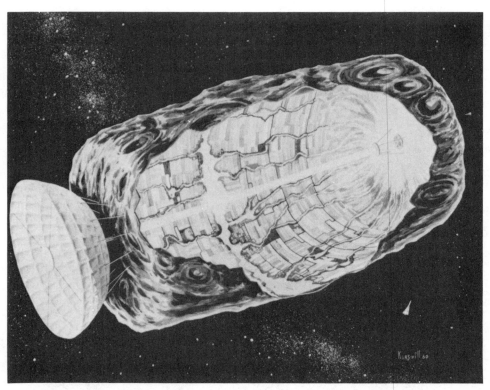

Figure 2-5. Dandridge M. Cole forecast working space colonies such as this in the early 1960s *(Martin Marietta)*

practically indestructible and immortal and will continue to grow and reproduce indefinitely. They could well represent the end product of the evolution of life.

We still see today copies and representations of Roy Scarfo's drawing of Dan Cole's hollow-asteroid space colony with its mass driver or catapult on its side. One cannot look at any of the modern illustrations of life inside a huge hollow space colony without comparing it to Scarfo's original painting of the view inside Dan Cole's hollow asteroid colony.

But Dan Cole was ten years ahead of his time. We hadn't yet landed a man on the Moon, an important step because it led to the cliché: "If we can put a man on the Moon, we can do anything, so why can't we . . . ?"

Gerard K. O'Neill

The next step took place when a high-energy nuclear physicist, Dr. Gerard K. O'Neill, proposed a study project to one of his MIT classes: Is a large space habitat technically and economically feasible? He discovered it was.

O'Neill already had considerable stature in the scientific community because of his invention of the storage ring for high-energy particle accelerators. He also became disturbed at what he read in the infamous book *The Limits to Growth*, by Dennis Meadows and others. He spoke of his concern to Dr. Hans M. Mark, then director of NASA's Ames Research Center in Sunnyvale,

California. O'Neill's initial paper on space colonies appeared in a publication no less prestigious than *Physics Today*, and Hans Mark found the meager funds necessary to pay for a "summer study" on space colonies at NASA Ames.

By combining the ideas of predecessors, thinking big, knowing how to present the concept to the world of science and to the media, and by accepting the necessary role of the charismatic leader as von Braun had done (von Braun was at that time in the waning years of his life and dying of cancer), Dr. Gerard O'Neill sparked a wave of space advocacy. The time was ripe for O'Neill, and he seized the opportunity. He sparked the 1970s grass-roots space advocacy movement.

Selling the Risky Concept

However, in common with all of the previous space pioneers, O'Neill was also somewhat ahead of his time. While there are very few people who doubt or question space colonization as an inevitable thing to come, O'Neill ran up against much the same barrier as other pioneers. I'd sat on platform panels with von Braun, Romick, and Cole while we attempted to convince engineers (who are eminently practical technologists) and politicians (who are eminently practical psychologists) that space colonization was a worthwhile endeavor that

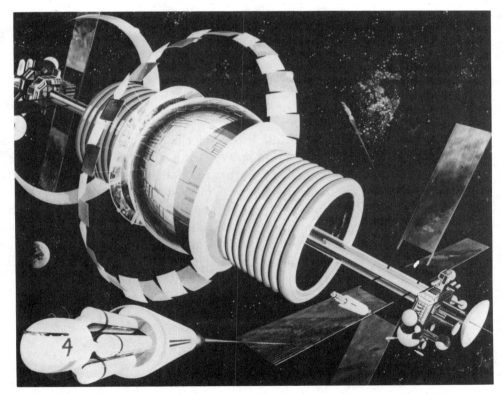

In the 1970s, Princeton nuclear physicist Dr. Gerard K. O'Neill sparked yet another wave of space interest with his concepts for large space colonies located at the Lagrangian points in the orbit of the moon. *(NASA)*

should be backed in spite of high risks because the payoff was enormous and would occur sooner than anticipated. I gave testimony with Dr. O'Neill before both Senate and House congressional committees with the same result. Therefore, if history can tell us anything (and it should), it's telling us that engineers can't and won't sell big engineering projects they think may be *slightly* beyond their well-understood state of the art. Engineers are not paid to take risks but to design bridges that stand and buildings that don't collapse.

The past is also telling us (if we read the history of the transcontinental railroad, the Panama Canal, and the airlines) that we shouldn't proceed with the belief that the government is going to colonize space because the government hasn't, isn't, and won't. The government cannot even run an adequate postal "service" or a passenger railroad. NASA has been told by highly paid consulting firms and NASA high officials admit that NASA cannot run the Space Shuttle as a space transportation system because NASA doesn't have the sort of people to do it. Nor do they have the necessary operating policies or procedures; and they are often prevented from taking correct or proper action by government regulations intended to control other government agencies.

1980s NASA Goals

NASA's goals for the 1980s have been carefully elucidated by President Ronald Reagan. They include a permanent manned presence in space. NASA has received the necessary approval to proceed with a space station. It is a very rudimentary space station serviced by the Space Shuttle. There's a good possibility that we may see it operational in orbit by 1991.

Red Star in Orbit

Meanwhile, the Soviet Union has already established a permanent manned presence in space with a series of space stations called Salyut, which, even though smaller than Skylab, have been launched into low earth orbit from the Baikonyr Cosmodrome at Tyuratam since 1971. Salyut is a two-man long-duration space habitat that has been visited by both male and female cosmonauts from the USSR and several allied countries. Soviet cosmonauts have lived in space for periods up to six months, and Cosmonaut Valeriy Ryumin made three missions to Salyuts and amassed a total time in space of more than a year. The Soviet Union has flown unmanned test models of a winged space shuttle, which is smaller than the NASA Space Shuttle.

Other Manned Space Programs

The French have plans for a small winged shuttle vehicle called Hermes that would be lofted into earth orbit with a cargo of seven people or their equivalent cargo weight by a planned advanced version of the Eurospace Ariane launch vehicle.

30

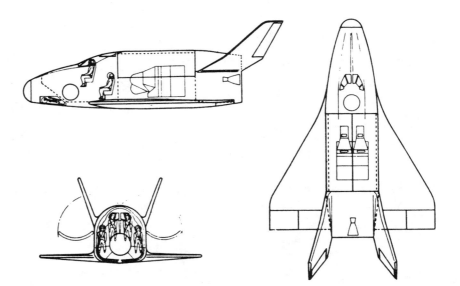

Figure 2-6. The French "Hermes" manned-space-shuttle concept that uses the European "Ariane" rocket booster

The People's Republic of China already has the technical capability and the launch hardware to orbit people and may well do so within ten years.

The Japanese are following their classical procedure of refusing to reinvent the square wheel by borrowing, licensing, or copying technology. By licensing the successful American "Delta" design, they've already developed the capability for orbiting large umanned satellites. The Japanese can put people in orbit anytime they put their minds to it.

Summary

We're on our way with a great legacy behind us. We've been thinking about space habitation for a long time. Many people have come up with many clever and elegant concepts. Billions of dollars have been spent in the development of manned space vehicles, most of them derived from military ballistic missile hardware or requirements. The Soviet Union has a permanent manned presence in space. Americans have the capability to get it but may not if government funding is depended upon in the future as it has been in the past. Free enterprise may step in to follow government exploration and development if space habitation follows the patterns of exploration and migration that occurred on past frontiers.

The only big question that now remains is: What language will space colonists speak?

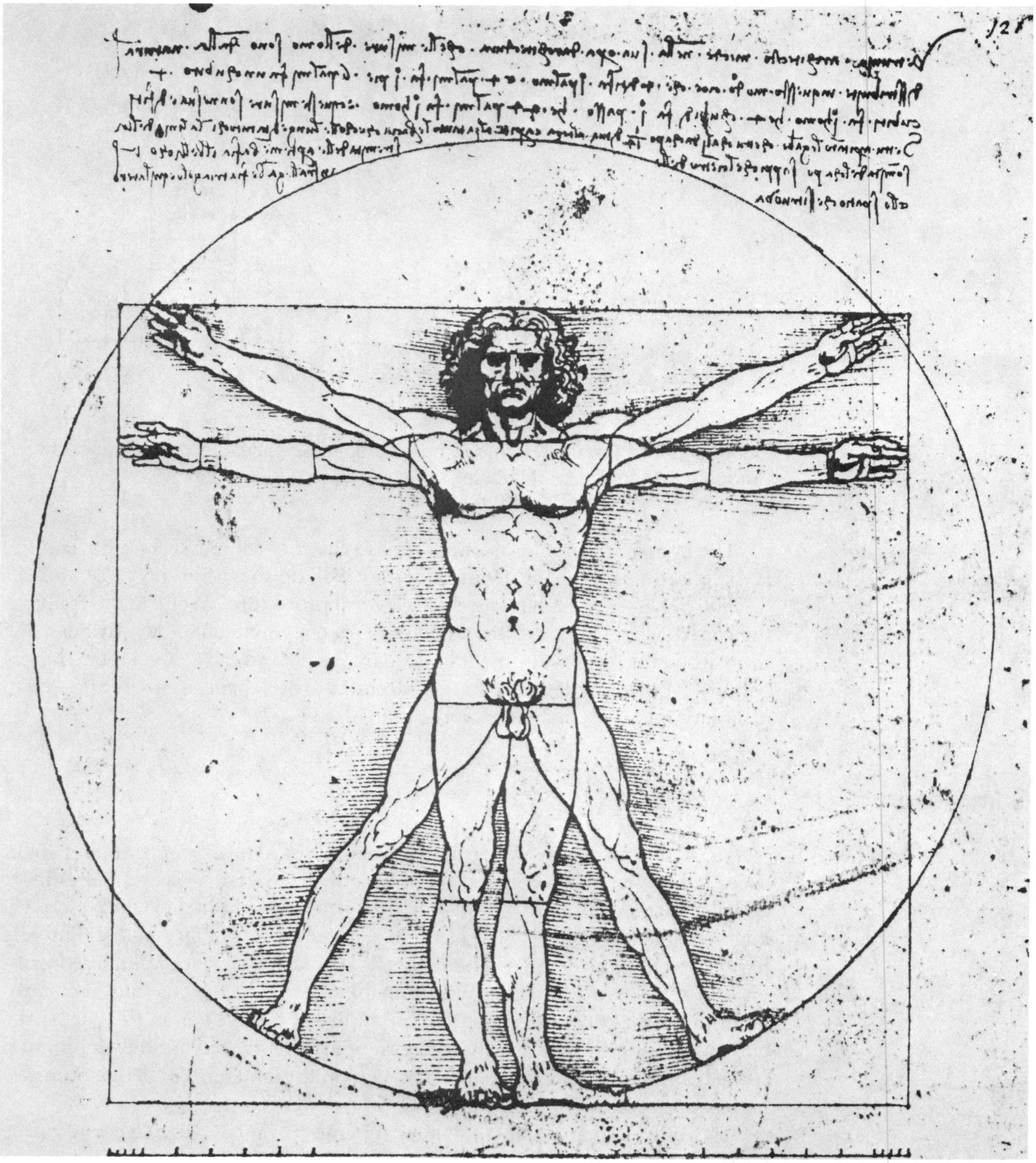

Figure 3-1. A human being can be looked upon as a machine, but is more than a machine. In going into space, we must have an accurate concept of what a human being really is. *(Leonardo da Vinci)*

3

What Is a Human Being?

In going into space, living in space, or designing and building devices for space travel and habitation, never lose sight of the fact that you are a human being and that other people in space are also human beings. You may think differently because you're pioneering a new frontier. But physically and mentally, you're no different from people on Earth.

In the process of inhabiting space, you must not only know the physical characteristics of the space environment but also something about yourself and the other people who are living and working there. A space habitat designed, built, and operated for dolphins, chimpanzees, or goldfish would be quite different in most respects from one designed for people. And people probably wouldn't be able to live in those habitats.

In dealing with anything, it's always good to know something about it. So we must answer the question: What is a human being?

Why Define a Human Being?

This may seem to be a sophomoric or even impractical exercise. After all, we are who we are, right? And we should know who and what we are, right?

Yes, but not quite yes.

For far too long, we've conducted human affairs on the basis of folk legends, fairy tales, old wives' notions, and perhaps an overinflated egotistical notion of who and what we are.

Progress in the human sciences—*not* the humanities, which have, unfortunately, fallen far behind—has been truly outstanding in the twentieth century. This should come as no surprise to readers who have been trying to keep up with things. We now know a great deal about a human being, thanks to progress in anthropology, biochemistry, neurology, psychology, and to some extent the slowly emerging proto-science of sociology.

The Schwartzberg Test

This last statement will undoubtedly raise the hackles of sociologists or those who believe that sociology is a real science. But let's see if we can define what we mean by "science," and then see if the shoe fits.

Science is universally defined as an organized body of knowledge.

The late Harry L. Schwartzberg of RCA observed that "the validity of a science is its ability to predict." This is based upon the assumption that if the past performance is known and enough experiments have been conducted with a reasonably large statistical universe, then the future reactions of a system to a given input can be accurately predicted within a known spread of statistical variation.

There is a very high probability that the sun will rise at New York City tomorrow morning at precisely the time given in *The American Ephemeris and Nautical Almanac.* There may be a 50 percent probability of rain in San Diego, depending upon how well the meteorologists have guessed that the weather systems will follow the patterns of movement similar systems have exhibited in the past. There is a high probability that Joe Smith at the corner bar will become intoxicated if he consumes six ounces of ninety-proof whiskey in one hour's time, no matter how well he holds his liquor. A public-opinion poll conducted with 1,250 people in a selected random sample of the American population will show results that would be duplicated within three percentage points if the poll were taken of the entire population.

These are examples of scientific predictions having varying degrees of validity.

Scientific Measuring Sticks

Dr. William O. Davis wrote in 1962: "Fundamentally, science must be a series of successive approximations to reality."

But in any science—and particularly in the human sciences—the ability to predict the effect of a given cause can be distorted by the philosophical, ideological, political, or personal beliefs or whims of the scientists involved in predicting the outcome. It is more difficult for these human factors to enter into the prediction if reasonably precise and accurate measurements can be and are made. Although *no* measurement is 100 percent accurate, it can be accurate enough for practical use. A lot of buildings are put together very well with a carpenter's rule, but that same measuring tool is useless to a micro-machinist or to an astronomer.

To achieve validity, measurements must be made. Alfred, Lord Kelvin, the famous British scientist, put it this way in 1886: "I often say that when you can measure something and express it in numbers, you know something about it. But when you cannot measure it, when you cannot express it in numbers, your knowledge is of a meager and unsatisfactory kind; it may be the beginning of knowledge, but you have scarcely, in your thoughts, progressed to the level of science, regardless of what the matter may be."

Therefore, in spite of the fact that many of the human sciences are still a bit "fuzzy," and even though the sociologists have started to make some measurements (Question: Do they yet know *what* to measure?), we still have enough data in hand at this time to make a reasonably accurate and valid definition of what a human being is so that we can properly ensure human survival in the new environment of space.

Defining a Human Being

It has to be a generalized definition, however, because every human being is different. There is an enormous variation of physical characteristics possible within the genetic structure of the forty-six paired chromosomes of an individual's "genome" and the extremely wide variety of environmental factors to which a given individual can be exposed and which shape development.

But we can start with a generalized definition. It will undoubtedly be modified and expanded as the years go by, just as it has in the past. But right now, we can legitimately ask the question, "What is a human being?" And we can give an equally legitimate answer:

A human being is a highly evolved bipedal, bisexual, unspecialized mammalian vertebrate with six unique special features, some of which are shared with other species of apes and monkeys but whose combination in *Homo sapiens* is unique among all animals.

These six features are:

- an erect posture

- freely movable arms and hands

- sharply focusing stereoscopic color-discriminating eyes

- a large brain capable of judgment and fine perception

- the power of speech

- less physiological and psychological differentiation between the sexes than in other species.

Each of these requires elaboration, especially if we're going to determine what a human being is so that we may provide properly for both human individuals and groups in space (and eventually here on Earth itself).

35

The Erect Posture

The erect posture is unique to human beings and was probably a survival asset that permitted early man-apes to gain the "high ground" of observation when looking out over the broad savannahs that became their hunting areas in prehistoric times.

But the evolution of the existing human skeleton from that of four-footed mammals is probably as much of a botched engineering job as the mammalian, reptilian, or amphibian skeletal structure is. The skeletal structure of vertebrates was originally evolved for water animals who exist in a world of balanced equilibrium between gravitational and buoyant forces. Gravitational forces still affect the internal organs of water animals. The skeleton merely serves as an anchor point for muscles and tendons, not as a structure that supports the entire mass of the animal against gravitation.

Mammalian skeletal structure is therefore full of weaknesses, some of which are shared by human beings:

The vertebral backbone is not a good structure from which to suspend an animal, especially a four-footed mammal. Humans aren't the only ones who have back problems; they're common with horses and dogs, too. The human backbone probably does a better job supporting the body as a column than a dog's back does supporting the canine abdomen as a beam.

The ball-and-socket hip joint of mammals—not just humans—is a structure with incredibly high strains involved because forces in the hipbone must be carried around a 90-degree turn just below the ball joint. Dogs suffer from hip problems; so do human beings.

The human rib cage is not as well designed for support as it is in horizontal animals. It requires an extremely complex and unsatisfactory network of muscles to keep the viscera in place even under normal gravitational forces.

However, the survival value of being able to stand and run erect overcame the structural shortcomings of the human skeletal structure and, in common with the other six features of a human being, is intimately interrelated with the other five.

Movable Limbs

The one great advantage of the erect posture of a human being lies in the freedom of movement of the forelimbs and hands that it permits. Although apes may occasionally stand erect, they don't move in an erect manner. Human beings share only with birds the fact that the forelimbs have evolved into different forms and are used differently from the hind limbs.

Because human beings stand and move erect, the human foot differs greatly from the hind feet of other animals, including the apes. The human foot has a big toe that points forward for walking, not sideward for grasping; the human foot cannot grasp as can an ape's foot.

And, in common with the birds, human beings must teach their young to move. In all other animals, the knowledge of how to move is preprogrammed into the brain and nervous system before birth, and the newly born animal can

36

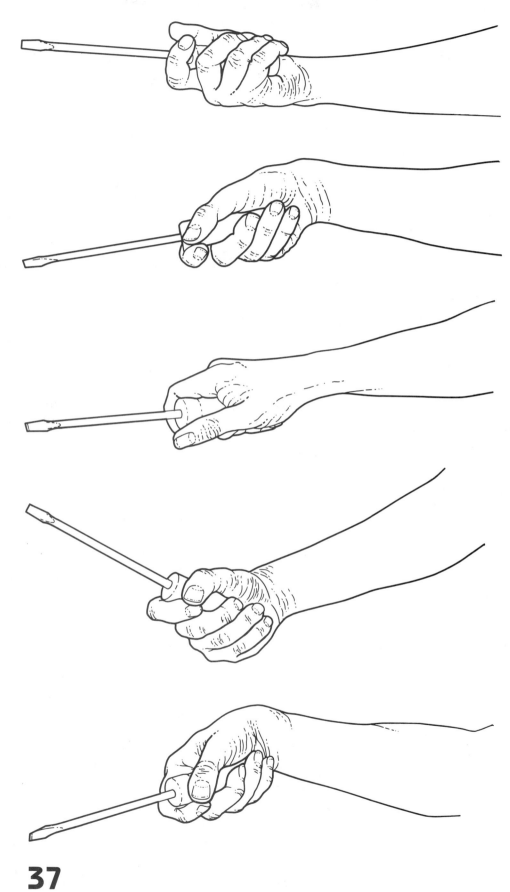

Figure 3-2. The human wrist is a marvelous device, with degrees of freedom in movement not possessed by the wrist of any other animal. *(Art by Sternbach)*

immediately move. Birds must teach their chicks to fly, and humans must teach their children to walk upright.

Walking erect freed human arms and hands from locomotion and permitted the further development of the human hand for manipulation. The human hand itself is unique among terrestrial animals. Although many apes possess an opposable thumb like that of a human being, none possesses the ability to move each individual finger separately, and none possesses the human wrist joint system.

The human wrist may be bent so that the hand is at right angles to the forearm. In this position, the hand may be rotated almost 180 degrees. While doing this and in any position in between, each individual finger may be moved. When the forearm is held horizontally, the palm of the hand may be turned up or down. No other animal on Earth has this sort of wrist joint.

The human hands and arms may be rotated in two nearly overlapping spheres from the shoulder, providing a human being with a large volume about the body in which the hands and fingers can manipulate objects.

The tip of each finger is a dense mass of tactile and kinesthetic receptors, providing a human being with a sense of touch perhaps unrivaled in the animal kingdom with the possible exception of the dolphins.

A large portion of the human brain is devoted to the task of sensing and controlling the human hand. More brain volume is dedicated to the thumb than to the entire visceral cavity.

Human beings therefore have two outstanding manipulators that are unmatched in their ability to feel the surrounding environment and to handle tools. Although humans may not have senses of smell and hearing that are as acute as those of some other animal species, human beings need not bow to any other species when it comes to the sense of touch.

Sharp-focusing Stereoscopic Eyes

In spite of the fact that some bird species are believed to have more acute vision, one of the unique human features is stereoscopic, sharp-focusing, color-discriminating eyes.

Like the apes, a human has two eyes set in the front of the skull with overlapping fields of vision. Some other mammalian species have eyes set partly on the sides of the skull, while amphibians and fish have eyes with no overlapping visual fields whatsoever. This overlapping of visual fields permits stereo vision, which gives humans outstanding depth perception while manipulating tools or materials.

The human eye also possesses an outstanding ability to "accommodate" or change the shape of its lens to focus on objects closely held in the hand. (As humans grow older, the stiffening of the tissues of the lens inhibits the ability to focus on nearby objects.)

The retina or visual receptor of the human eye is also excellent when it comes to color discrimination. The retina contains color-sensitive visual receptors known as cones (from the general shape), which are tightly packed in the retina around the prime focus of the eye system. The majority of the retina is

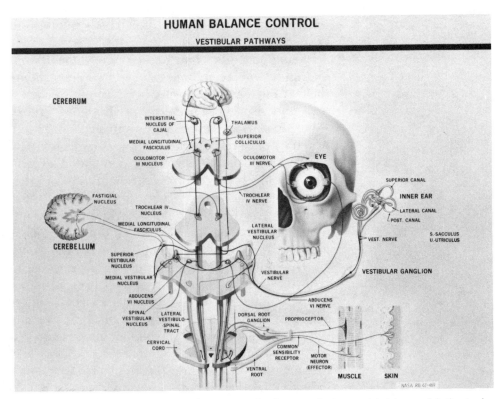

HUMAN BALANCE CONTROL

VESTIBULAR PATHWAYS

Figure 3-3. The balance-control system of a human being, a highly sophisticated assembly of organic sensors, computers, and end effectors *(NASA)*

made up of rod-shaped visual sensors which are not sensitive to color but are instead much more sensitive to light. At high light levels, cone vision over-whelms rod vision, and humans therefore see in color and in great detail; at low light levels, rod vision predominates, color washes out to various shades of gray, and the eye becomes primarily sensitive to motion. Therefore, although colored detail is sensed near the primary axis of vision, the remainder of the eye's visual field is primarily used to detect motion. Thus, the human visual system performs the dual survival role of being able to sense in detail and in color whether or not the fruit held in the hand is indeed ripe while the peripheral visual field is constantly alert to sense the motion of something lurking in the bushes.

The human visual system may still be evolving. The Greeks had no word for the color green, and the past fifty years have seen the development of new colors and extremely delicate shades of colors that were never used in the past, if the art of painting can be relied upon as accurately duplicating what the artist actually saw.

The Large, Complex Brain

The factors of erect posture, highly sensitive and manipulative hands, and sharp-focusing binocular color vision required the development of a large brain

39

in terms of brain weight per unit of body weight. And it required the evolution of specialized parts of the brain.

Standing erect is an exceedingly difficult thing to do and requires an enormous volume of brain to handle the constant inputs from tactile sensors in the feet and legs, the sensory input from the balance organs located behind the ears, and visual cues from the environment. Human children will pull themselves erect as apes do, but only human children can be taught to walk upright. And they must be taught; the ability to walk is not inherited. This feat requires a large brain volume devoted to perception and association.

The same brain requirements hold true with regard to the ability to use the human hands and fingers. Human children spend a great deal of time learning how to use their hands and, again, they must be taught to manipulate an eating utensil, which is basically the first tool a child must master. As the human child grows and matures, the manipulation of other tools must be taught.

The ability to play the piano, paint a picture, fly an airplane, or use a screwdriver must be taught, and all these abilities require a brain that possesses the additional feature of being able to exercise judgment and make decisions, which also must be taught. Some recent work in psychology indicates that the left side of the human brain is dominant in learning these things and in controlling the pertinent parts of the human body in the manipulation of the outside world, while the right side of the brain has the ability to use judgment and discretion.

These four unique human features all require a brain large in size with relation to the body. A newborn human child is almost a brain with a small body attached. And, unlike other mammalian species, the human child's big brain must be taught the things that make the organism into a human being.

The Power of Speech

This created the need for the fifth unique human factor: the power of speech. Other apes possess a limited power to communicate by sounds. Gibbons have a vocabulary of nine separate sounds such as imperatives like, "Stop, danger!" or "Stay away from my mate!" Animals in general can verbally warn or command, but they cannot verbally transmit past experience or future plans to others of their species. In this regard, human beings are unique.

The development of language probably came from the need to teach human children what to do—to walk, to hunt, to make tools, to use tools, etc. The human uniqueness in locomotion and transport, perception, and ability to handle tools and materials is not preprogrammed into the human brain as similar but less-developed factors are in animals. The human brain evolved from a preprogrammed, instinctually driven control organ into one that is programmed to *learn*. The power of speech and the development of language evolved from this fact. And from the power of speech, the power of deductive logic and the act of creation grew because human beings think in words.

When you study another human language, you should also study the culture in which the language is used. Every human language is a *complete* communication system which mirrors as well as satisfies the cultural needs of those who speak it. It also mirrors the cultural way of thinking. For example, in a culture where little change takes place, the structure and grammar of the language becomes quite complex, formalized, and stiff. As a culture becomes more complex and changes rapidly, the structure of the language grows simpler to permit rapid lingual change. Many language teachers, writers, lexicographers, and grammarians decry the current world tendency to speak mixed languages where words are borrowed outright from other languages; and most resist the vulgarizing of popular speech and writing. The greatest resistance is mounted against the use of jargon. However, lingual mixing, vulgarizing, and jargon are age-old factors that permit a spoken and written language to continue to be a means of communicating new concepts, ideas, principles, and techniques.

The human development of a large brain capable of handling the unique developments of erect posture and bipedal locomotion, free-moving hands and arms capable of precise manipulation, fine binocular color vision, and the power of speech for communication, all permitted human beings to achieve mastery of Planet Earth. However, one additional factor has recently come to light that undoubtedly provides great assistance in doing this.

Differentiation of the Sexes

Human males and females show less sexual differentiation than any other animal. That is, there are fewer strong physical and psychological differences between men and women than in any other species. Both men and women share numerous secondary sexual characteristics. Both sexes exhibit amazingly similar physical strengths and endurance to a degree not believed possible even a decade ago. Although the human female must physically bear the human child, either or both parents are equipped to raise that child to be an adult human being.

This is probably not a new thing in the human race. Once, human beings were a small, rare breed apart; there may have been fewer than a thousand of them on Earth. They couldn't afford the luxury of the division of labor in the hunt because there were too few of them. When they began to till the soil, they were still few in number and half their children died of natural causes within five years of birth. Those males and females who could work together in teams survived; those who were weaker than their mates didn't make it. It was only when human beings conquered all other species and became masters of the planet that their numbers began to increase and they started expanding into every ecological niche they could manage to occupy. Then the division of labor became a necessity, and the first such division was between male and female.

Because the good times were few and scarcity was the general lot of humanity for most of the time on most of the planet, this cultural sexual dimorphism was artificially encouraged. It's still strong in those cultures living in scarcity. In the modern advanced cultures where abundance is the general rule, sexual differentiation is disappearing and the original teamwork and equality between human sexes is reappearing.

Whereas the human mastery of transportation, communication, and mate-

41

rial handling gave the species mastery of the planet, the unspecialized nature of the sexual team may be the key to mastery of space because it may be the critical factor in the development of successful extraterrestrial cultures.

The Cultural Animal

The physical development of the human species has of necessity created human beings as cultural creatures. The ability to pass on to others the experience of the past and the potentials for the future made possible by the power of speech is itself the root of culture.

The late historian Will Durant wrote in the introduction to his monumental series, The Story of Civilization:

"Civilizations are the generations of the racial soul. As family-rearing, and then writing, bound the generations together, handing down the lore of the dying to the young, so print and commerce and a thousand ways of communication may bind the civilizations together, and preserve for future cultures all that is of value for them in our own. Let us, before we die, gather up our heritage, and offer it to our children."

But what are the essential elements of a human culture, a group of individual human beings drawn together and interacting between themselves within the group and in a unified manner with other human groups? What are the environmental and social elements that underlie all successful cultures of the past and must therefore provide the foundations for people living together in space?

Again, the least element in any human institution or culture is the human individual, regardless of the ideology of the group.

Modern human cultures appear to be complex because of the many interrelating institutions in which an individual participates. Anthropologists have found that all the basic elements of institutions are present in the simpler hunting cultures from which all of our modern ones evolved. Therefore, it's possible to get an excellent and highly detailed picture of the cultural and environmental requirements of a human being by referring to these simpler cultures. The extremely close relationship with modern living will immediately become evident.

The Open Environment

Human beings apparently evolved on Earth and are physically attuned to the specific characteristics of the terrestrial environment. This environment must be taken with them into space for reasons that will be discussed later in detail. But beyond the physical characteristics of the environment are those characteristics which are more subtle and subjective: the psychological and social environmental characteristics.

Man—a contraction of the word "human" and not a term applying exclusively to the male of the species—evolved as a hunting mammal who is happiest if the cultural environment contains space to move around in. These surroundings should be as "natural" as possible or give the impression of being natural. They should include other animals, if possible, and green growing things. This is an evident trend in modern cultures, where animal pets have taken the place of barnyard animals and thriving indoor plant and outdoor gardening and landscaping industries have emerged.

The attempt to achieve this environmental goal exists in all human cultures even if, in some, it becomes a reality only for a few. Actually, it's a requirement for all mammals because scientific experimentation has revealed that when too many rats are placed together in too small a space, they begin to exhibit bizarre behavior of a nonsurvival type; human beings are the only animals to do this to themselves voluntarily, and everywhere it exists there are social problems that seem to be insoluble. And perhaps they are under those environmental circumstances, people being what they basically are.

The Basis for Work and Play

Humans need physical exercise and recreation. In spite of the fact that Europeans and Asians have learned how to live together in the restrictive space of cities, these cities have parks and places to play. Our modern medical technology has rediscovered the physiological and psychological need for exercise (which some medical doctors and physical-fitness enthusiasts have extended to the extremes, as always happens).

The physical and mental problems faced by people who are, physically and psychologically, hunters and working at tasks in a sedentary environment are now well known but perhaps not completely understood yet.

And human beings need recreation. "Play" is not a unique activity among humans because animals engage in playful, nonhurtful competition and individual interaction either as practice for the real thing or as diversion. Again, human enthusiasts have taken recreation to its extremes, an action that may only be an experimental attempt to discover the limits of play.

Play is the foundation for the arts, for the willing suspension of disbelief that permits the socially acceptable acting out of fantasies, and for the creative act that underlies both science and art. This exists in all human cultures.

A human being needs something to do, a job that is at times dangerous, that at all times requires the use of more brain than brawn, and that sharpens the wits by presenting something new and different which appears to be (or is believed to be) soluble and doable. Hunting is an occupation that includes all of these things. Farming doesn't. Nor does work on an assembly line in a factory building, the dehumanizing legacy that has come down from the early days of the First Industrial Revolution in England. Although many people long for a life of leisure, those who have been able to live such a life soon discover that they become bored or ill, either physically or mentally.

Social Interaction

A human being needs a small group of people with whom he can interact on a face-to-face basis informally and without fear of being perceived as a threat or of being threatened. This group may include family, close friends, or even a formally organized club. If these groups become too large, the individual feels a loss of control over his personal relationship with the group and a threat of possible coercion by the group leaders over whom he can exert no personal influence.

As social groups become large, formalized group behavior develops because it becomes important to know who's in charge of what, what an individual's responses must be in any given situation, and what responses are to be expected from others.

Because humans evolved as superior hunters who could cooperate in bringing down a kill much larger than any of them, people know that cooperation is absolutely essential when large and dangerous jobs are tackled. But the hunting experience that bred the human race also placed a premium on isolated individual action, perhaps contrary to plan, when and where necessary for personal survival or for the successful completion of the task.

The hunter's mind also wants to make his personal experience available to the group to ensure success. Therefore, people want the freedom to express opinions, present data from experience, and form their own opinions based on their evaluation of the situation or of the individuals who lead or propose to lead the group.

When people don't do what's expected of them or when they perform destructive acts against other people in the group, human beings must indicate the wrongness of such behavior and attempt to prevent the recurrence of it by inflicting punishment that ranges from *lex talionis*—an eye for an eye—to banishment from the group. An organized system or code of behavior must be developed that includes provisions for handling behavior deviating from the code. Thus a legal system that embraces the group's concepts of justice emerges. Such systems are most effective when the majority of the members of the group participate in the process and may become totally ineffective if such power gravitates into the hands of a few.

Usefulness in Space Living

All of this and more is applicable to living and working in space, the development of the new extraterrestrial civilization. Even in the early primitive days of its development in the twenty-first century, this hard-won knowledge of who and what human beings are must be applied to what is done in space.

And although in space it's possible to start afresh without the ghosts of the past to haunt us, we can't afford to make the same mistakes all over again. It's extremely dangerous and expensive to reinvent the wheel, especially in the dangerous environment of space. Human beings will have to make use of all of their unique characteristics, and especially the ability to communicate the lessons of the past.

Keep all of this in mind when you face the inevitable physical and social problems of space habitation. You, too, are human. This description is you.

Figure 4-1. Mankind's journey into space began when people first left the surface of the Earth in 1783 to travel into the atmosphere with primitive hot-air balloons.

4 Atmospheres

Staying alive, healthy, and happy in space requires many different things and depends on many technologies. But basically your life and health in space depend upon your own Earth-evolved physiology. Before you can successfully work with the nuts and bolts of life-support systems, you need to learn what they have to support in the first place. Some life-support systems are more critical than others. And the safety rules now in use and those developed out of hard experience in the years to come will depend upon these basic facts about yourself.

The first area of life support that must be approached involves atmospheres.

Basing this discussion on the human problems involved with flight in the Earth's atmosphere may appear at first glance to have little relationship with living in space. However, the conquest of space was preceded by the conquest of the air during which we learned how important were the physiological effects of zero pressure, reduced pressure, and even overpressure—factors you'll have to live with every second of your life in space.

Without an atmosphere around you at adequate pressure and with proper composition, you'll die. Quickly. Insofar as your body is concerned, space begins at the Earth's surface itself. Because of the unique characteristics of the Earth's atmosphere, you can experience conditions approaching those in space without leaving the atmosphere. As you ascend upward through the atmosphere to space, conditions become more spacelike, leading a pioneer space scientist, Dr. Hubertus Strughold, to propose the concept of "partial space equivalency" as a function of altitude above the Earth's surface.

The space between the worlds is devoid of any sensible atmosphere whatsoever and, as far as you're concerned, can be considered as a vacuum.

The gaseous atmosphere of the planet Earth appears to be unique in the solar system. Other planets and large satellites possess atmospheres, but these don't have the necessary combination of proper pressure and composition to permit you to live without what is currently termed "life-support equipment."

The Earth's Atmosphere

Until two hundred years ago, people thought that the atmosphere was everywhere, making it possible to fly to the Moon in an open basket lifted by trained birds or even a balloon. The atmosphere was thought to permeate the entire universe. The personal experiences and reactions of people who lived or traveled in mountainous regions were ignored or misinterpreted because of this "pervasive atmosphere" belief. It wasn't until scientists began to measure various atmospheric factors and explorers started to ascend to high altitudes in balloons that the true nature of this important environmental factor was discovered.

The first scientist to measure something about the Earth's atmosphere was Evangelista Torricelli (1608–1647), who was a colleague of Galileo and a mathematician and physicist in Florence, Italy. In 1643 he invented the mercury barometer (Figure 4–2) and concluded that it was the weight of the atmosphere that sustained a column of mercury inside the inverted closed tube. This was quite contrary to the belief of the time, the doctrine of Aristotle that "nature abhors a vacuum." In 1648 Florin Périer carried a "torricellian tube" up the mountain Puy de Dôme and found that the height of the mercury column was less at the summit.

The pressure of the Earth's atmosphere at sea level will support a column of mercury 760 millimeters (29.92 inches) high. The amount of pressure required to do this is 14.696 pounds per square inch. This pressure is created by the weight of trillions of molecules of air in a column 1 inch square extending from the surface of the Earth at sea level upward. The air molecules are pulled toward the Earth and are thus given weight by the Earth's gravity.

There are minor daily and seasonal variations in atmospheric pressure at any point on the Earth's surface that are caused by something that Earth has and space does not: weather. Localized pressure changes of as much as 2 inches of mercury or 1 pound per square inch are not uncommon. In the middle of a hurricane, pressure can drop as low as 26 inches of mercury, or about 2 pounds per square inch less than normal.

Figure 4-2. The mercury barometer, or "torricellian tube," in which the column of mercury in the tube is held up by the air pressure on the surface of the mercury in the lower bowl or reservoir *(Art by Sternbach)*

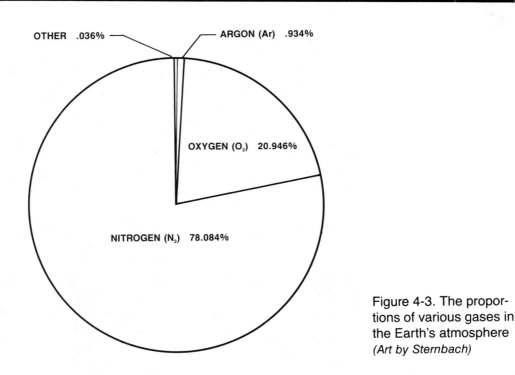

OTHER .036%

ARGON (Ar) .934%

OXYGEN (O$_2$) 20.946%

NITROGEN (N$_2$) 78.084%

Figure 4-3. The proportions of various gases in the Earth's atmosphere *(Art by Sternbach)*

As one ascends into the atmosphere or climbs a mountain, the column of atmosphere grows shorter and thus the pressure exerted by that air column becomes less. Near the surface of the Earth, the atmospheric pressure decreases roughly 1 inch (25 millimeters) of mercury or about ½ pound per square inch with every 1,000 feet of altitude.

The Earth's atmosphere is a mixture of gases having approximately the following makeup (numbers rounded off):

Nitrogen 78.09%

Oxygen 20.95%

Argon 0.93%

Carbon dioxide 0.03%

Other gases such as neon, helium, krypton, hydrogen, xenon, ozone, and radon make up the minute fraction of a percent remaining.

This composition doesn't vary worldwide except occasionally in the vicinity of large cities and industrial complexes where more carbon dioxide and particulate matter may be concentrated. However, there's a greater percentage of carbon dioxide over the oceans of the southern hemisphere than over any large industrial metroplex.

Although pressure decreases as altitude increases, the composition of the atmosphere remains reasonably the same below about 50,000 feet, or 15 kilometers.

The density of the atmosphere—i.e., its weight per unit volume—under Standard Temperature and Pressure or STP conditions (a temperature of 59°

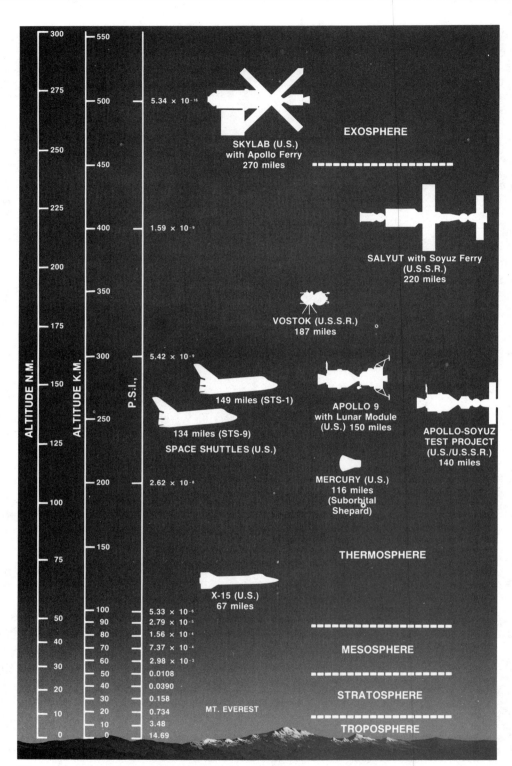

Figure 4-4. In order to go to and from space, spacecraft and their passengers must pass through the Earth's atmosphere, whose regions have generally been divided into layers by scientists. Human beings have evolved at the bottom of this atmosphere. *(Art by Sternbach)*

Fahrenheit and a pressure of 14.696 pounds per square inch) is 0.76475 pounds (mass) per cubic foot. As altitude increases, air density decreases along with pressure. This means that at high altitudes there are fewer molecules in a given volume of air.

The "sea level" or normal characteristics of the atmosphere are important to know and remember because (a) they represent the conditions that your body considers "normal," and (b) they're nearly the conditions you'll have to maintain around you in order to live in space.

Oxygen

You can't stay alive for more than a few seconds without being surrounded by a medium that exerts pressure upon your body's external surface and contains a gaseous component, the element oxygen, at a minimum gas pressure. Because your body has evolved in the unique gaseous atmosphere of Earth and uses breathing to obtain the oxygen needed to support life, this pressure and composition must closely resemble that of normal sea-level conditions on Earth.

This critical environmental factor should be something that's well understood by everyone, but it's actually both taken for granted and generally misunderstood. Until only recently in biological time we've always lived, traveled, and worked with the Earth's atmosphere around us. Only in the past 200 years have we had the capability to leave the surface of the Earth and travel in the atmosphere.

And only within the last two centuries have we learned what the prime purpose of breathing is.

The fact that the atmosphere changes with altitude provided the first clues concerning how the human body is affected by surrounding pressure and how strongly dependent upon physiology the operation of the human body is. Dr. Strughold's "partial space equivalency" was experienced by people within a very few years after the Montgolfier brothers made their initial manned balloon flights near Paris in 1783.

Your body is a heat engine that consumes the fuels of carbohydrates, fats, and proteins from food by the chemical process of oxidation, which requires the presence of oxygen. You obtain this oxygen from the surrounding atmosphere by the process of respiration, utilizing your lungs. Not until 1660 was the true function of your lungs in respiration discovered by Marcello Malpighi (1628–1694) in Bologna, Italy. The fact that your lungs provide your body with oxygen and remove the products of cellular oxidation, carbon dioxide and water, was discovered by the French physiologist Claude Bernard (1813–1878).

At rest, you consume approximately a half-pint of oxygen gas a minute at STP but you can require as much as 8 pints per minute when running.

Inhaled oxygen molecules from the air must pass across the alveolar membrane of your lungs into the capillaries that carry the blood to your lungs. Then the oxygen molecules must attach themselves to your red blood corpuscles. To

51

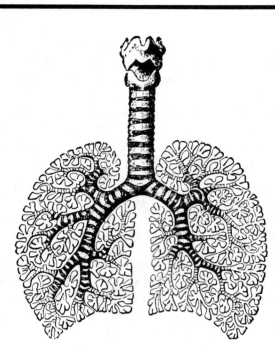

Figure 4-5. In order for oxygen to reach the bloodstream and then be transported to body tissue where it's needed for metabolism, oxygen molecules must have enough partial pressure to cross the membranes of the alveolar sacs in the lungs and then replace carbon dioxide being carried by red blood corpuscles.

do this, they must have enough energy to dislodge from the red corpuscles the carbon dioxide molecules that are oxidation products carried there from other parts of your body. This energy is expressed in terms of pressure.

Pressure

John Dalton (1766–1844) discovered that, in a mixture of gases, each component gas behaves as if it alone occupied the total volume of the gas mixture and exerts a pressure equivalent to its percentage of the mixture. Known as Dalton's Law of Partial Pressures, this can be restated: "The sum of the partial pressures of each constituent gas in a mixture equals the total pressure of the mixture."

And it was the French physiologist Paul Bert who pointed out in 1878 that it's the partial pressure that determines the physiological effects of a gas.

Thus a gas mixture such as air containing 20 percent oxygen and with a total pressure of, for example, 10 pounds per square inch (psi) would have an oxygen partial pressure of 10×0.20 or 2.0 psi.

When applied to the atmosphere at STP, Dalton's Law tells you that you're used to living in an atmosphere with oxygen at a pressure of 3.08 pounds per square inch.

Extremely careful pressure measurements made inside human lungs have shown that the pressure of carbon dioxide on the blood side of the alveolar membrane is approximately 40 millimeters of mercury, or 0.7735 pounds per square inch. This is about one-fourth as much as the oxygen partial pressure in the Earth's atmosphere at STP. As long as the oxygen partial pressure in your lungs is greater than the partial pressure of the carbon dioxide, oxygen will flow inward across the membrane to replace the carbon dioxide, and the carbon

52

dioxide will flow outward so it can be expelled from your body with your exhaled breath.

These physical facts provide the foundation for the most critical and hazardous dangers to human beings who fly at high altitudes or live in space: hypoxia, anoxia, and hyperventilation.

Lack of Oxygen

Hypoxia is a condition of oxygen shortage in your body, while *anoxia* is a condition where insufficient oxygen is available to sustain your life. They're two aspects of the same environmental factor and differ only in degree. The condition of hypoxia can exist for hours or more, while anoxia can be fatal in a matter of minutes. But because of the importance of pressure, hypoxia and anoxia cannot be considered apart from *hypobaria* and *abaria* (reduced pressure and lack of pressure).

Aviators flying at high altitudes in the atmosphere were the first to experience hypoxia. The first hypoxic situations were reported during World War I, when the fragile airplanes of the time began to achieve altitudes of about 15,000 feet above sea level.

The cause of hypoxia is related to partial pressure. The partial pressure of oxygen becomes less with increasing altitude, and density also decreases, meaning that there are fewer oxygen molecules per cubic foot of air. The combination of lower oxygen partial pressure plus lower oxygen density means that when you inhale surrounding or "ambient" air, there are fewer oxygen molecules to pass through the alveolar membranes of your lungs and into the bloodstream; and because of reduced partial pressure they enter your lungs with less energy available.

Hypoxia, whether encountered while flying or in space, is perhaps the most insidious of all dangers because if you're a victim of hypoxia, you don't realize it's happening.

The first portion of your body to suffer from lack of oxygen is the most important one: your brain. Although your brain comprises only about 2 percent of your body weight, it requires 25 percent of the oxygen you inhale by respiration. And within your brain, the "higher functions" or most recent evolutionary portions are the parts initially affected by hypoxia.

The first symptoms of hypoxia are insidiously pleasant and resemble mild alcohol intoxication:

- Your normal self-critical ability is dulled.

- Your capability to exercise judgment by comparing and analyzing alternatives is greatly impaired.

- Your brain's ability to correlate different sensory inputs disappears.

- Your memory becomes elusive.

- Important matters no longer seem important because hypoxia has a tranquilizing effect.

53

Your motor control and coordination are next to suffer. You become clumsy without being aware of it. Hypoxia brings on feelings of well-being, drowsiness, nonchalance, and a false sense of security.

The last thing you'll think necessary is oxygen.

At night or under conditions of reduced illumination, your vision suffers and becomes dim, again without your being aware that it's happening.

Many of the same hypoxic sensations are experienced regularly and voluntarily by persons who smoke tobacco products. Nicotine has a much greater affinity for red corpuscles than for oxygen, and a smoker deliberately subjects

26,000 Feet — Chamber Test

INSTRUCTIONS:

Log Distances and Leg Times were given and Pilot was asked to Compute Ground Speeds and Total Times.

Oxygen Mask Removed after 4th Entry. Note the errors and Deterioration of Legibility.

1 Misread Computer

2 Used Wrong Computer Scales

3 Error in Addition

4 Number Omitted

5 Loss of Consciousness

GROUND SPEED	DISTANCE N.M.	TIME	
		LEG	TOTAL
153	28	11	11
180	63	21	32
174	110	38	70
180	36	12	82
174	43	15	107
212	76	28	35
\	80	29	

Figure 4-6. The effects of hypoxia. This is what happened when a pilot in a pressure chamber was asked to make and write down calculations when the chamber was at various equivalent altitudes. As altitude increased, the mistakes became more obvious, until unconsciousness set in. *(Federal Aviation Administration)*

his brain to a mild form of hypoxia as well as carbon dioxide imbalance. Airplane pilots who smoke are known to have less tolerance to hypoxia than nonsmokers. By smoking a single cigarette, a smoker automatically puts his body at an altitude of 8,000 to 9,000 feet because of the carbon monoxide in the inhaled smoke.

In addition, tobacco smoke contains a higher concentration of radioactive radon gas. This gas is present in minute quantities in the Earth's atmosphere but not to any harmful level because all of us have evolved in that atmosphere. Radon produces alpha particles during its radioactive decay, and this radiation has been directly linked to cancer. Thus, even though you may not be a smoker yourself, if you inhale tobacco smoke you'll be getting a shot of radioactive radon gas.

A person who smokes is a high risk in space not only because of these factors but also because of the effect of tobacco smoke tars and other ingredients upon components of the life-support system. Make a quick visit to any local airline

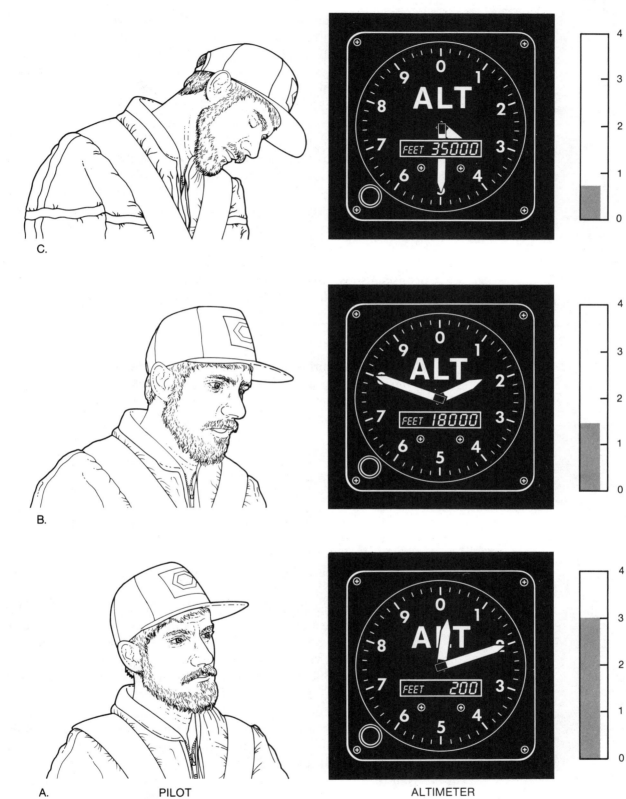

| PILOT | ALTIMETER |

Figure 4-7. The effects of hypoxia vary according to altitude, atmospheric pressure, and partial pressure of oxygen *(Art by Sternbach)*

maintenance facility and see the brown goo that must be removed from fans, blowers, filters, and other parts of the cabin air-conditioning system of a jet airliner. In space facilities, this sort of deliberate gumming up of the works can't be tolerated because if the air system stops working, so do the people who depend upon that system.

As hypoxia gets worse, you lose the ability to balance or orient yourself and get dizzy. There may be a tingling of your skin. A dull headache may begin, but it's usually only partly perceived at this stage and you'll dismiss it as a mild annoyance because of the advanced state of tranquilization created by hypoxia. Your heart rate increases as your body attempts to pump more blood through your lungs and brain to rebalance the oxygen and carbon dioxide situation in your body. Your lips and the skin under your fingernails turn blue. Your field of vision narrows until peripheral or cone vision disappears, leaving only rod vision, which itself slowly becomes blurred as the nerves that control the ability of your eyes to accommodate or focus begin to stop working for lack of oxygen.

Unconsciousness finally occurs if the oxygen partial pressure continues to decrease or drops to a level where insufficient oxygen is available to your brain.

Anoxia

At that point, the condition of anoxia is predominant. You may go into shock or convulsions caused by carbon dioxide poisoning, which upsets the acid-base balance in your body, creating acidosis. This condition alone may kill you long before the oxygen level in your body drops to a point where your brain stops working altogether.

Preventing Hypoxia and Anoxia

The first data concering hypoxia came from aviators. If you fly at high altitudes, safety regulations now require that you begin using an on-board supply of supplementary oxygen fed via an oxygen mask over your nose and mouth (or a nasal cannula in your nostrils) when flying at altitudes above 12,500 feet. At that altitude, the oxygen partial pressure is 1.92 psi or about 62 percent of its sea-level value.

However, hypoxia can begin working at altitudes as low as 8,000 feet (oxygen partial pressure of about 2.3 psi or 75 percent that at sea level) with people who are elderly, overweight, or heavy cigarette smokers. Since hypoxia drastically affects cone vision, which is essential for night vision, pilots at night often begin using supplemental on-board oxygen at altitudes of 8,000 feet.

Thus the lower limit for the onset of hypoxic effects can be set at an oxygen partial pressure of about 2.0 psi.

Severe hypoxia occurs at 18,000 feet altitude where the total ambient pressure, and therefore the oxygen partial pressure, is half that at sea level (Figure 4–9). At an altitude of 18,000 feet, you're physiologically halfway to space. Half the atmosphere is below 18,000 feet. And the partial pressure of oxygen has dropped to 1.54 psi.

It's possible to breathe supplemental oxygen through a mask up to an

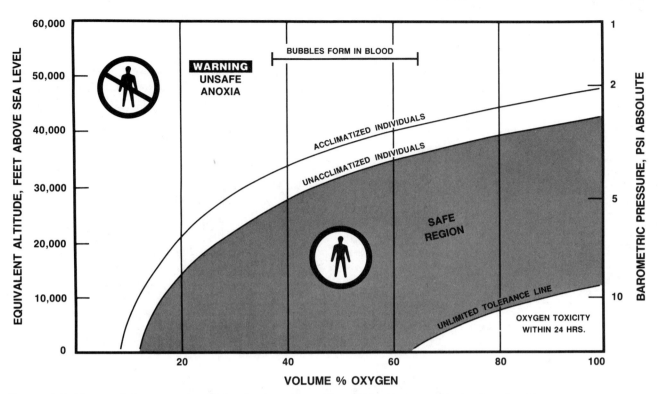

Figure 4-8. Human tolerances to atmospheric composition and pressures *(Art by Sternbach)*

altitude of approximately 35,000 feet, which is equivalent to an oxygen partial pressure of 0.7735 psi. This is equal to the carbon dioxide partial pressure in your lungs. Thus there's no pressure or energy differential that will permit the oxygen to replace the carbon dioxide in the blood in your lungs; so oxygen must be supplied to your lungs at a pressure greater than ambient. This is called "pressure breathing." Oxygen under pressure inflates your lungs and can therefore displace the carbon dioxide there; but you must then forcibly exhale against the pressure. This is the opposite action to normal breathing, and it becomes extremely fatiguing in a short period of time.

At altitudes in excess of 45,000 feet, the pressure differential between the oxygen in the pressure-breathing system and the total ambient pressure becomes too great, and it becomes impossible for you to exhale.

Partial Pressures of Atmospheric Gases as a Function of Ambient Pressure

Gas	Percentage of Air	Ambient Pressure (psi)		
		14.7 (sea level)	7.35 (18,000')	3.46 (35,000')
Nitrogen	78.09%	11.476 psi	5.738 psi	2.711 psi
Oxygen	20.95%	3.079 psi	1.539 psi	0.727 psi
Carbon dioxide	0.03%	0.004 psi	0.002 psi	0.001 psi

Figure 4-9.   *Note:* Alveolar pressure of carbon dioxide in lungs is approximately 0.7735 psi.

Pressure breathing with an oxygen mask can't be used in space where there's zero ambient pressure. Pressure must be applied to your body either by means of surrounding it with a gaseous atmosphere or by applying physical pressure to your entire external body surface.

The full-pressure suit helmet of the Apollo space suit, which completely surrounds a person with an Earthlike atmosphere *(NASA)*

There's another factor involved here besides the partial pressure of oxygen: As atmospheric pressure decreases, the boiling temperature of water also decreases. Anyone who's tried to cook a meal at high altitudes in the Rocky Mountains quickly becomes aware that boiling water isn't as hot as it would be at sea level. At an elevation of 18,000 feet, where half the atmosphere is below you and the ambient pressure is 7.32 psi, the boiling point of water is depressed to about 180°F. At an altitude of approximately 65,000 feet where the atmospheric pressure is about 1 psi, the boiling point of water is depressed to 98.6°F, which is blood temperature. Technically, without the application of additional pressure to prevent it, your blood will boil at pressures below this, which means that, if you're suddenly exposed to the vacuum of space, you face the scientific possibility of having the blood in your alveolar capillaries turn into vapor. Actually, this doesn't happen in reality because of your skin, as will be discussed later.

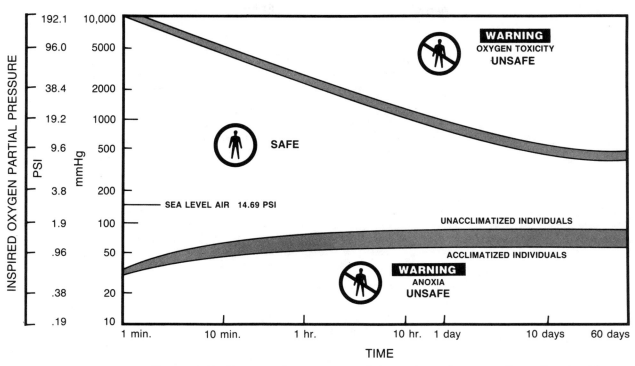

Figure 4-10. Human tolerance to partial pressure of oxygen *(Art by Sternbach)*

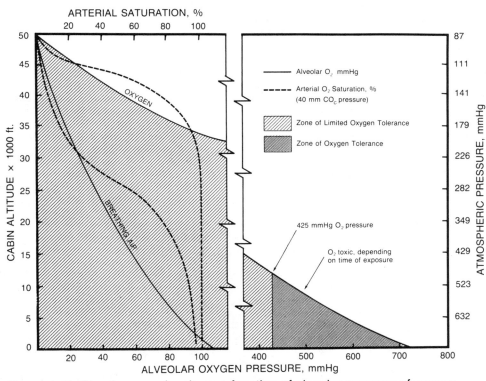

Figure 4-11. Blood oxygen levels as a function of alveolar pressure of oxygen
(Art by Sternbach)

59

It's also possible and just as hazardous for you to get too much oxygen. Rapid and deep breathing, called "hyperventilation," can occur voluntarily, but it usually takes place under stress. It's easy to experience a mild form of hyperventilation. It's a common children's game. Just begin breathing rapidly and deeply. The symptoms of hyperventilation will become very evident very quickly. The age-old survival mechanism of our hunter ancestors goes into action, pouring the substance adrenaline into your bloodstream and triggering

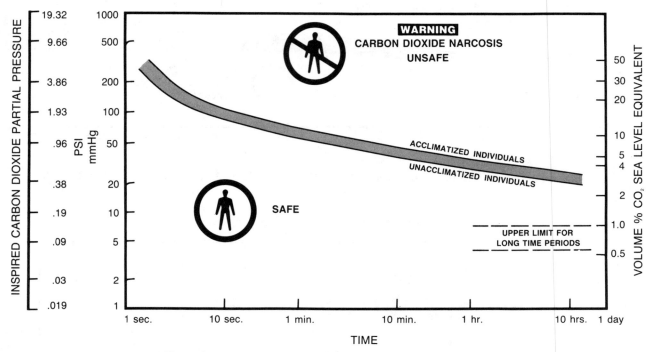

Figure 4-12. Human tolerance to partial pressure of carbon dioxide *(Art by Sternbach)*

the "fight or flee" syndrome that's a biological part of each of us. Adrenaline puts your body in an emergency condition that includes rapid, deep breathing in anticipation of immediate and extensive action.

But hyperventilation can flush too much carbon dioxide out of your blood and lead to oxygen toxicity. The presence of carbon dioxide in your blood causes your blood serum to be slightly acidic, and removing too much carbon dioxide therefore changes the chemical balance of your blood. This produces dizziness, tingling of the fingers and toes, a sensation of body heat, increased heart rate, blurring of vision, muscle spasms, and, if carried on long enough, unconsciousness. The symptoms are very similar to those of hypoxia in many ways, although it's brought on by precisely the opposite environmental conditions.

Unlike hypoxia and anoxia, hyperventilation can be overcome quite simply if you recognize the condition. First, check to be certain that it's indeed hyperventilation and not hypoxia brought about by a failure of your life-support system, your pressure suit, or your pressure cabin. If everything appears to be normal in that department and the condition is verified as hyperventilation, make a conscious effort to slow down your breathing rate and depth. The easiest way to do this is to *talk*. Sing. Count aloud. Carry on a conversation even if there's nobody around to talk with (nobody will know). Normally paced conversation tends to slow down and stabilize a rapid breathing rate, and so does singing. Normal breathing is an immediate cure for hyperventilation.

Nitrogen and the Bends

Deep-sea divers and free divers using scuba equipment have encountered the opposite conditions because they continually work under *increased* environmental pressure depending upon the depth to which they dive. They've encountered a phenomenon known as nitrogen narcosis or "rapture of the deep." While you may not run into this because it occurs under pressures of 50 psi or more, you may encounter other problems with nitrogen.

As with most other environmental factors, either too much or too little nitrogen causes trouble. For example, shortly before the second flight of the NASA Space Shuttle in 1981, a launchpad worker was killed because he inadvertently entered a compartment in the tail of the Orbiter that was filled with nitrogen gas being used to flush any toxic gases from the compartment. Exposure to a pure nitrogen atmosphere is almost instantly fatal, even though nitrogen is a harmless, nontoxic gas that makes up more than 70 percent of the normal atmosphere you breathe. Several lungfuls of nitrogen displace all the oxygen in your lungs, leaving your blood loaded with carbon dioxide. Death is painless and rapid. The launchpad worker, as with other individuals who've succumbed to a 100 percent nitrogen atmosphere, never knew what hit him.

Nitrogen is the preferred diluting gas for all space facilities because it's the major diluent gas in the Earth's atmosphere, and you're attuned to it. Therefore, you may encounter badly imbalanced nitrogen-rich atmospheres, another hazard involved in space pioneering.

Although nitrogen is preferred over other gases such as helium (long used by deep ocean hard-suit divers) because of the relative availability of nitrogen versus that of helium, the use of nitrogen as the major diluent creates the potential for other major problems in space, problems that have been known for decades here on Earth by divers, underwater caisson workers, and aviators.

The major concern, even in the Space Shuttle, is that of evolved gas or "the bends."

Like oxygen and carbon dioxide, nitrogen also crosses the alveolar membrane of your lungs and goes into solution in your blood serum. When your body is subjected to a loss of pressure, or hypobaric condition, over a period of time lasting from seconds to several hours, depending upon the actual pressure change in psi, the nitrogen dissolved in your blood comes out of solution and forms tiny bubbles.

61

This is the same physical mechanism that takes place when a bottle of soda water, beer, or carbonated soft drink is opened. While the content of the bottle is under pressure, the carbon dioxide that makes the fizz remains in solution; the instant the pressure is released by opening the bottle, the carbon dioxide comes out of solution as tiny bubbles of gas.

When this same thing happens to the nitrogen in your depressurized body, the nitrogen gas bubbles tend to congregate at the joints of your arms and legs, where their presence creates almost unendurable pain, "the bends."

However, bends are rarely observed with changes in pressure of 7 psi or less.

There's also a solution to preventing the bends: time.

Divers ascending from the depths of the ocean come up in a series of short lifts, spending a certain amount of time at each depth as they come up, in order to let the dissolved nitrogen come out of their blood slowly. The number of steps and the time spent at each one is determined from a set of divers' tables.

Pilots of high-performance high-altitude military aircraft spend thirty minutes prior to takeoff breathing 100 percent oxygen through a mask in order to flush most of the nitrogen out of their bodies and thus prevent the bends.

Space Shuttle crew members, however, were initially required to spend three hours in their space suits breathing pure oxygen to flush the nitrogen out of their bodies before venturing outside the cabin into the vacuum of space. This precaution was taken to prevent the bends in case the space suit develops a leak and pressure is lost. The Space Shuttle space suits operated at a pressure of 5 psi while the cabin pressure was 14.7 psi. Bends were possible. But there was no absolute requirement to spend three hours getting rid of evolved nitrogen since about 75 percent of the evolved nitrogen can be flushed out by a 100 percent oxygen prebreathing process, but NASA aeromedical experts wanted to be absolutely certain no Shuttle crew member could possibly encounter the bends since there was no way, under certain emergency conditions, that they could have been assured of getting them back to normal pressure quickly.

You may encounter bends in space if you're in a pressurized habitat or module that suffers a rapid decompression. If that happens, however, the bends will be the most painful part of your immediate problem but not the most critical. Your immediate concern and action must be to get back into pressure at once. Whether or not you encounter the bends in a space suit will depend totally upon the design of the suit, its operating pressure, and its safety measures.

Bends may turn out to be an important warning of a loss of pressure and the onset of hypoxia that would accompany it.

Carbon Monoxide Poisoning

Although you might not think of it, carbon monoxide is one of the greatest of all dangers in the space environment. Carbon monoxide on Earth usually comes from the exhaust of an internal combustion engine. But carbon monoxide results from incomplete combustion of any sort.

Other than the exhaust of an internal combustion engine, the most common source of carbon monoxide on Earth is cigarette smoke, which contains about 3 percent carbon monoxide. A pack-a-day smoker goes around with 4 percent to 8 percent of his blood saturated with this gas, which is roughly half the lethal concentration.

In space, carbon monoxide may come from electrical equipment that gets too hot, from improper food cooking techniques, or from any of a number of sources that we can't yet imagine but that some of you early space pioneers will discover because you're carrying out the prime function of pioneering.

Carbon monoxide is an odorless, colorless gas that has an affinity for red blood cells 200 times greater than that of oxygen. Carbon monoxide poisoning therefore is a form of hypoxia and anoxia because it keeps your red blood corpuscles from carrying oxygen to the tissues of your body. The onset of carbon monoxide poisoning is the same as and just as insidious as hypoxia—i.e., you won't know it's happening until perhaps it's too late.

There are a number of carbon monoxide detectors on the market. Some of them are chemical in nature and require no power for operation. If you're smart and want to stay alive, you'll see to it that carbon monoxide detectors are installed and working wherever needed.

Summary

To recapitulate your human requirements for pressure and atmospheric composition so that you can breathe:

1. You require gaseous oxygen at a pressure of about 3 pounds per square inch.

2. You exhale carbon dioxide, which is a product of cellular combustion or oxidation.

3. A careful balance between oxygen and carbon dioxide in your breathing air must be maintained so that your body can maintain the proper chemical balance in your bloodstream.

4. Nitrogen is a diluent gas in the normal Earth's atmosphere, but its presence can create severe problems with evolved gas if the atmospheric pressure is decreased.

5. An excess of any type of atmospheric gas can be hazardous.

6. Too much or too little atmospheric pressure can be hazardous.

Life-support systems can handle all of these requirements if they're working right. But knowledge of the *why* will be vitally important to you if there's a life-support system malfunction or when you find yourself in any of a large number of hazardous emergency situations in space.

Figure 5-1. Although the thermometer was originally invented by Galileo in 1592, no accurate measurement of temperature was possible until the early 1700s when Fahrenheit in Germany developed the mercury thermometer and R. A. F. de Réaumur invented the alcohol thermometer.

5

Keeping Cool

Having the proper atmosphere around you in space is only part of the business of staying alive there. As on Earth, you can be surrounded by an atmosphere with pressure and composition that's perfect for your needs, yet the temperature of that atmosphere can be too hot or too cold for you to live in.

Temperature control in space, however, is more than just heat-energy management. As on Earth, there's also the matter of humidity.

Earth's Thermal Environment

The Earth is also unique among the bodies of the solar system because it's the only celestial body we know of on which liquid water exists in the free state because of its atmosphere, its gravitational field, which provides that atmosphere with adequate pressure, and the temperature on its surface. This combination of conditions allows the Earth's atmosphere to hold water in a vapor state.

Under normal Earth conditions, water vapor is invisible. However, when the atmosphere becomes saturated with water vapor and can hold no more, any additional water suspended in the air condenses into microscopic droplets. When this happens, so much water is available that the trillions of microscopic water droplets can be seen as a whitish haze or as clouds.

Earth isn't the only planet to possess clouds of water droplets, but no other planet has such an abundance of water. Almost 70 percent of its surface is at all times covered with cloud layers.

Because you evolved on this unusual planet with its water-laden atmosphere, the effects of temperature can't be discussed without also considering relative humidity.

Relative Humidity

"Relative humidity" is a term often used with little real knowledge of what it means and how it relates to your comfort.

Any volume or parcel of air is something like a transparent sponge. It will hold a maximum amount of water in vapor form. When it won't hold any more, it's said to be saturated. The total amount of water that a parcel of air can hold depends upon its temperature and pressure and is termed its "absolute humidity." This is expressed in terms of weight of water per given volume of air—i.e., grams per cubic centimeter, ounces per cubic foot, etc.

For many reasons, an air parcel is rarely completely saturated with water vapor. The actual amount of water vapor in it at any time can be expressed conveniently in terms of the percentage of its absolute humidity. This is known as the relative humidity.

The warmer the air parcel and the greater its pressure, the more water vapor it can hold. If the air parcel is cooled, its water vapor will condense to form fog, haze, clouds, and, in extremes, water that condenses into large droplets that fall as rain, freezing rain, sleet, or water crystals called snow.

Hot air will hold more water vapor than cold air because water vaporizes more easily at higher temperatures. Thus, an air parcel at 90°F holds more water vapor than a parcel of equal volume and pressure at 30°F.

When an air parcel is cooled to the temperature at which it becomes saturated (also known as the "dew point"), the excess water vapor condenses in the form of fog, clouds, or precipitation. This can be easily demonstrated on any hot summer day when the cold surface of a glass of ice water forms condensation on its surface. The cold surface causes the air to cool, lowering its temperature to the saturation point in terms of the amount of water vapor in the surrounding air. The water vapor condenses on the glass.

As warm, humid air rises from the Earth's surface, it cools and the water vapor in it also condenses.

The process is reversible. When water goes from the liquid state to the vapor state, it absorbs heat and cools any object with which it is in contact. When water evaporates into air, the air is cooled along with the surface from which the water was evaporated. Although water can't evaporate into air that is already saturated, the drier a parcel of air is—i.e., the lower its relative humidity—the easier it is for water to evaporate into it.

This is the principle behind the air-conditioning unit known as the "swamp cooler," which is used in the American Southwest. The hot, dry desert air is drawn through fibrous pads wetted with water that evaporates into the dry air, lowering its temperature by as much as 30°F but also increasing its relative humidity.

Because of the cooling effects of evaporation, relative humidity is extremely important to your comfort and strongly affects your temperature tolerance.

Body-Temperature Control

Your body is extremely sensitive to heat and works best between very narrow limits of temperature. You must maintain your internal temperature of 98.6°F, especially in critical locations such as your brain. Vital organs of a highly specialized and critical nature such as your brain *must* remain at a steady temperature or they'll stop working. Some parts of your body can get much colder or hotter than 98.6°F without permanent damage; but if your limbs and extremities get too cold, they can become frozen.

Like a home heating system, your body has a built-in temperature sensing and control system. When you get too hot, temperature sensing nerves, acting like the thermostat in a house, signal your brain of this condition. Your brain then automatically opens the pores of your skin to release perspiration. This perspiration evaporates into the air to cool your body.

If you get too cold, your pores close and your skin puckers up into goose pimples or "duck bumps." This is an instinctual reaction to cold that's inherited from the time when our remote ancestors were covered with a thick coat of fur. They fluffed up their hair to provide additional insulation against loss of body heat by trapping air between the hairs of their fur. These days, you have no fur to fluff up to provide a thick mat of insulation, but your body still thinks the fur is there and tries to fluff it up. This causes goose pimples. (These days, you can provide yourself with artificial fur by putting on a sweater or coat.)

Measurement of Relative Humidity

A twin thermometer called a "psychrometer" is needed to determine relative humidity and THI (see below). A psychrometer has one ordinary thermometer alongside an identical thermometer whose temperature bulb is covered with a wetted piece of cloth. The dry bulb measures temperature like any other thermometer, while the wetted bulb thermometer measures the cooling created by evaporating the water in the wick.

Relative humidity is determined by reading the wet-bulb and dry-bulb temperatures and consulting a complex psychometric chart such as the one shown in Figure 5–2.

The Temperature-Humidity Index

Because of your built-in cooling system, you must always know the temperature *and* the relative humidity when dealing with the effects of temperature. This has led to the use of an empirical number called the "discomfort

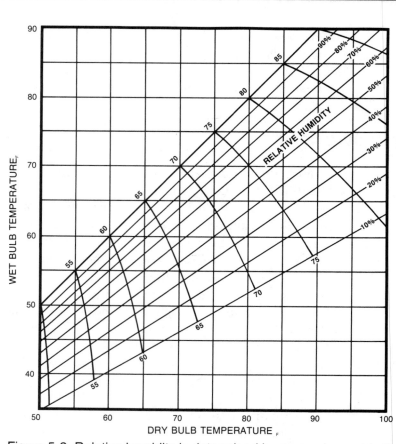

Figure 5-2. Relative humidity is determined by measuring wet-bulb and dry-bulb temperatures and referring to this "psychometric," or temperature-humidity chart. *(Art by Sternbach)*

index" or the Temperature-Humidity Index (THI), which is reported in the news media in connection with the daily weather during the summertime.

THI is calculated by adding the wet-bulb and dry-bulb temperatures together, multiplying the result by 0.4, and adding 15.

There's no real reason for these numbers. They are "empirical"—they've been developed by rule of thumb from experience.

A THI of 75 means that most people are suffering from discomfort; a THI of 80 or more produces acute discomfort for almost everyone.

Effects of High Temperatures

However, above certain temperatures relative humidity plays a minor role, if any, in terms of your comfort.

At a temperature of 95°F, you'll be uncomfortable at any humidity. When the temperature is over 102°F, you'll be acutely uncomfortable, regardless of the humidity, if you remain in this thermal environment without taking steps to control your body temperature.

In the deserts in a place like Phoenix, Arizona, summertime temperatures often soar as high as 115°F. This is *hot*, but it's bearable because the relative humidity in the desert climate ranges between 3 percent and 10 percent. You can remain reasonably comfortable in desert temperatures of 110°F or more because your perspiration evaporates very rapidly into the hot, dry air, thereby cooling your body.

However, in New York City or Houston, a temperature of only 85°F can make you *extremely* uncomfortable when the relative humidity is between 80 percent and 90 percent, which is usual and normal. You become soaked with perspiration that, because the air is already nearly saturated, doesn't evaporate quickly enough to cool your body and carry away your body heat fast enough. It isn't the ambient temperature that affects you as much as your inability to get rid of your own body heat.

Effects of Low Temperatures

Much the same effects are present in cold weather but are the result of somewhat different physical mechanisms.

In dry desert climates, winter cold seems milder, although the temperature often dives well below freezing and can stay there for days. The cold seems mild because of the low humidity. When air is dry, it can't hold as much heat as when it's wet with water vapor—and it can't absorb as much heat from your body. So the air can't cool your body as much or as quickly as wet air.

In a cold, damp climate, the winter weather seems to bite into you because the damp air can carry away more of your body heat faster than dry air.

Temperature and Heat

Temperature is not the same as heat.

Temperature should be thought of as the *speed* of molecules; the faster they move, the higher the temperature. Temperature is a measure of the degree of hotness.

Heat, on the other hand, is a general term meaning the quantity of thermal energy in a given amount of matter.

For example, a pound of water may contain the same amount of thermal energy as ten pounds of water, but its temperature will be higher because the same amount of thermal energy exists in *less* volume of the same material.

Heat-Transfer Methods

Heat is transferred from one piece of matter to another by one or a combination of three methods: conduction, convection, and radiation.

Conduction heating is transferral of thermal energy by direct contact. A pan on a stove gets warm because it's in contact with the hot surface of the stove. Heat flows from the hot stove to the cooler pan.

Convection heating is a form of conduction heating. If you put water in the pan on the stove, the water at the bottom of the pan is heated by conduction from the pan. Being a fluid, it grows less dense as it is heated and therefore rises from the bottom of the pan because it is lighter than the cooler water surrounding it. This allows cooler surrounding water to move in to be heated. The heated water while rising passes some of its heat to the water above it by conduction. Convection heating is, thus, conduction heating speeded up by a mixing process.

Radiation heating is something totally different. A person sitting in front of a sunlamp or basking in the summer sun on a beach isn't being warmed by conduction or convection. A lamp, the sun, or any warm body gives off heat in the form of infrared radiation. This radiation is the same as radio waves, microwaves, light, X rays, or other electromagnetic radiation except for its frequency. Infrared requires no medium for transmission. At the surface of a hot body, heat energy becomes infrared radiation, which travels away from the body in all directions at the speed of light. When this infrared radiation reaches another body, it's converted back into heat energy.

You can be warmed or cooled by a combination of these three methods.

On a cold day, if you sit on a radiator to warm up, you gain heat by conduction.

When you sit in a tub of water, you're being heated or cooled by convection, depending upon the water temperature.

An electric heater or the sun warms you by radiation.

Air-Movement Effects

The motion of air over your body has a great effect upon your comfort because of the principle of conduction and convection heating. Air motion provides forced convection.

On a hot, humid day, a breeze or the air movement created by a fan removes the saturated air next to your body, replacing it with fresh air that hasn't been saturated by water vapor from your perspiration.

On the other hand, a combination of low temperatures and air movement will make you feel colder than the actual temperature for the same reason and because of the same principle. The wind strips away the layer of air next to you even if you're wearing heavy clothing. This creates a wintertime problem known as "wind chill."

A table of wind-chill factors (Figure 5–3) has been empirically developed in the same manner as the THI charts. It indicates, for example, that a temperature of 20°F with a wind of 20 miles per hour will cause a body heat loss equivalent to that in −10°F with no wind; in other words, the wind makes 20° feel like −10°.

Human Thermal Tolerances

THI, wind chill, and other factors relating to the high and low limits of human temperature tolerance weren't known until the Department of Defense went

Cooling Power of Wind on Exposed Flesh Expressed as an Equivalent Temperature

Estimated wind speed (in mph)	Actual Thermometer Reading (F.)											
	50	40	30	20	10	0	– 10	– 20	– 30	– 40	– 50	– 60
	EQUIVALENT TEMPERATURE (F.)											
calm	50	40	30	20	10	0	– 10	– 20	– 30	– 40	– 50	– 60
5	48	37	27	16	6	– 5	– 15	– 26	– 36	– 47	– 57	– 68
10	40	28	16	4	– 9	– 24	– 33	– 46	– 58	– 70	– 83	– 95
15	36	22	9	– 5	– 18	– 32	– 45	– 58	– 72	– 85	– 99	– 112
20	32	18	4	– 10	– 25	– 39	– 53	– 67	– 82	– 96	– 110	– 124
25	30	16	0	– 15	– 29	– 44	– 59	– 74	– 88	– 104	– 118	– 133
30	28	13	– 2	– 18	– 33	– 48	– 63	– 79	– 94	– 109	– 125	– 140
35	27	11	– 4	– 19	– 35	– 51	– 67	– 82	– 98	– 113	– 129	– 145
40	26	10	– 6	– 21	– 37	– 53	– 69	– 85	– 100	– 116	– 132	– 148

Wind speeds greater than 40 mph have little added effect.	LITTLE DANGER (for properly clothed person) Maximum danger of false sense of security.		INCREASING DANGER Danger from freezing of exposed flesh.	GREAT DANGER		

Source: NAVMED Bulletin 5052-29

Figure 5-3. Wind-chill chart. The higher the air velocity, the lower the apparent temperature and the greater the "wind chill." Chart is valid not only for Earth, but also in space habitats. *(Art by Sternbach)*

to work on the problem. A soldier in the Arctic or a flier in the stratosphere must be clothed warmly and satisfactorily, or he will just sit around trying to keep himself warm instead of attending to his duties. A soldier can't be sent into the tropics or the desert in a fur coat and be expected to function. You can't perform at top efficiency if you're too hot or too cold. Therefore, to learn about the temperature tolerance of human beings, the armed services have made an enormous number of experiments concerning the effects of temperature, humidity, and ventilation using thousands of volunteers as experimental subjects. Data was taken with the volunteers fully clothed, optimally clothed for the temperature-humidity-wind environment, and unclothed.

The data from these experiments have been compiled over several decades. In analyzing the results, researchers have gained an increasing knowledge of human tolerance and reaction to what they call "thermal stress."

Comfort is complex and subjective. It's not a simple subject to deal with, as is shown by the complex "comfort chart" (Figure 5–4) that relates dry-bulb temperature, wet-bulb temperature, relative humidity, and air movement. Thermal stress occurs outside the normal zones of comfort.

If you get too cold, the chemical processes in your cells slow down and the cells eventually die. If your body liquids freeze, they'll crystallize and destroy the fine structures of the cells.

On the other hand, if you get too hot, you can literally burn yourself up because your body is a heat-generating system. If you can't get rid of the heat generated by organic combustion, the complex organic molecules inside your cells will break down or cease to function.

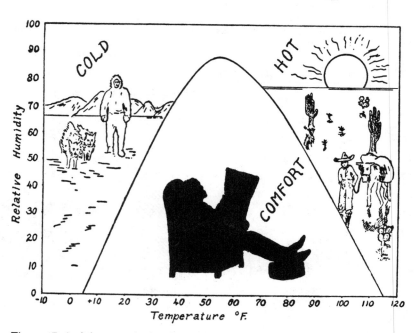

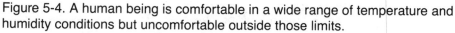

Figure 5-4. A human being is comfortable in a wide range of temperature and humidity conditions but uncomfortable outside those limits.

The Human Heat Engine

Your body is a heat engine that combines the oxygen from your lungs and the food from your digestive tract into carbon dioxide, water, and heat.

Heat is measured in calories. A calorie is the amount of heat required to raise one gram of water 1°C (or 0.02 ounces of water 1°F). This is not the same calorie as those you count when dieting; a nutritional Calorie (capitalized) is equal to 1,000 thermal calories.

At rest, your body gives off nearly 1,600 calories per minute. During active periods, your thermal output jumps to about 2,900 calories per minute. In comparison, your body gives off about as much heat as a 100-watt light bulb. It is thus understandable why a room full of people quickly grows warm just from the accumulating body heat of everyone there.

The total heat you create from the combustion of food is about 3,000 Calories per day, assuming a daily routine or schedule as shown in Figure 5–5.

While you're resting, about 78 percent of this heat is given off by normal heat-transfer processes such as conduction, convection, and radiation. Twenty-two percent is given off by evaporation of perspiration.

While working, you lose 45 percent of your heat by normal heat-transfer methods and 44 percent through perspiration. The remaining 11 percent goes toward the actual accomplishment of physical work.

Calculations based on these figures indicate that you're approximately 11 percent efficient, or about equal to the efficiency of a gasoline engine.

They also assume that you're in an optimum temperature-humidity environment—not too hot and not too cold.

Because of these tests on human volunteers made for military purposes,

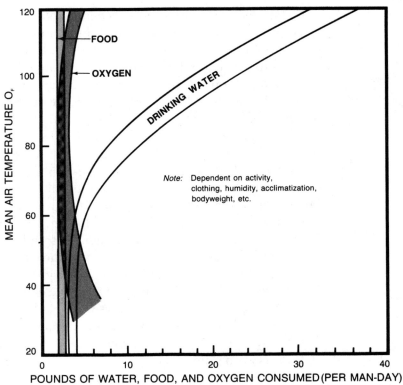

Figure 5-5. Human food and water needs as a function of temperature
(Art by Sternbach)

your temperature and humidity tolerances as well as the absolute limits of temperature that a human being can withstand are now well known and thoroughly documented. This is a classic example of swords beaten into plowshares: The test results have had an enormous influence upon the design of clothing for outdoor and sporting purposes.

Clothing

Proper clothing is absolutely mandatory for human survival anywhere—on Earth or in space.

As discussed in chapter 3, *Homo sapiens* has evolved as an unspecialized hunting species. Long, long ago your ancestors gave up the insulating protection of a hairy coat. You retain only remnants at critical locations on your body where body parts must maintain proper temperature under the worst environmental circumstances. Hair acts in conjunction with sweat glands in cooling your body; hair acts to increase the evaporative surface for perspiration under conditions of high ambient temperature while serving as an insulating layer to trap air under conditions of low ambient temperature.

However, your ancestors learned how to provide themselves with artificial coats by using the hair and skins of other animals. This invention permits quick adaptation to different ambient temperatures as well as providing protection against solar radiation. Rather than be trapped by the environment, your ancestors learned how to take a comfortable environment along with them anywhere they went. As a result, they conquered climate and were able to spread across the entire Earth from the equator to the poles.

And now you are doing the same thing in the voids of the solar system.

Clothing is one of the great and continually improving inventions of mankind. It has evolved locally into garb that will protect an Eskimo against the arctic climate as well as the Arab against the harsh desert environment. Clothing isn't just ethnic; it's pragmatic. It's been developed and perfected century by century by local people and is quickly adopted by other people who move there. The parka, the haik, the caftan, the muumuu, the overcoat, the cowboy hat, the greatcoat, the kepi, the turban, the T-shirt, Bermuda shorts, boots, the wool sweater, gloves, earmuffs, sweatbands, pith helmets, and long johns are all incredibly sophisticated low-technology products brought about by centuries of testing and improvement. The oldest industry and the first to be touched by the industrial revolutions was the clothing industry, along with its innovative testing area, the fashion industry. You take all this for granted on Earth because, so far as you're concerned, clothing has always been there. From diaper days on, you've always worn clothing.

Without clothing, the environmental range in which you could live comfortably wouldn't extend far beyond the dry, temperate prairie-steppes where clothing is worn primarily for protection against chafing rather than as portable environmental control.

Human Temperature-Humidity Tolerances

Figure 5–6 shows the temperature tolerances of optimally clothed people as a function of temperature and time. This data will hold true in space just as it does on Earth because it deals with the same subject: you, *Homo sapiens sapiens*, Model One.

Properly clothed, you can indefinitely withstand temperatures between approximately −30°F and 120°F. Note that the high end of the temperature tolerance depends upon the relative humidity.

Temperature tolerance, as indicated in Figure 5–6, is time-dependent. For shorter periods of time and if you're acclimatized, or used to it, you can tolerate a wider range of temperatures—from about −80°F to more than 140°F.

A temperature of 160°F is tolerable for about 30 minutes. You'd be capable of about 5 minutes of activity at a temperature of 200°F and 10 minutes of activity at a temperature of 150°F.

A temperature of 120°F is tolerable for about an hour but is considered to be above the range for continual physical or mental activity.

Actually, mental activities begin to slow down, errors begin to creep in, and complex performance begins to deteriorate at temperatures above 85°F, regardless of the humidity.

Physical labor begins to be extremely fatiguing at temperatures above 75°F.

On the low end of the temperature scale, physical stiffness of your arms and legs begins at 50°F. Cold injury to extremities is a function of time as shown in Figure 5–6.

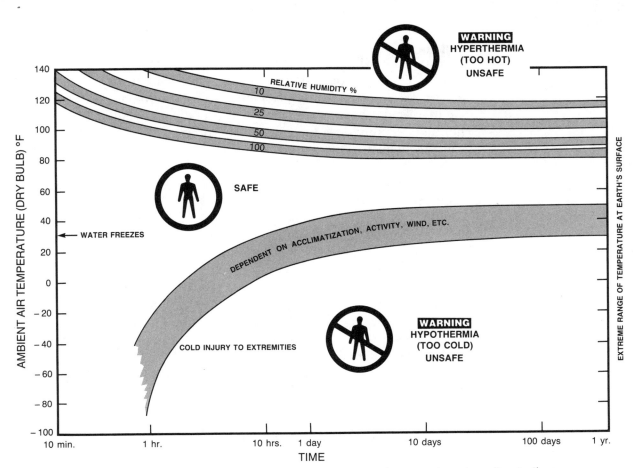

Figure 5-6. Human tolerance to temperatures varies according to time.
(Art by Sternbach)

Although some temperature-humidity tolerances are shown in Figure 5–6, a low humidity of 15 percent or less at any temperature can cause drying up of your external body fluids and the membranes of your nose and mouth. On the high end, humidities in excess of 90 percent are generally considered by most people to be intolerable.

The most comfortable temperature-humidity region for unclothed human beings lies between 65°F and 90°F with relative humidity between 30 percent and 40 percent and air-movement velocity between 30 and 50 feet per minute.

Beyond these limits lie the hazards of hyperthermia (too much heat) and hypothermia (too little heat).

Effects of High Temperatures

Hyperthermia includes a number of conditions all caused by the same thing: your body becoming too hot.

One of these conditions is heat prostration caused by failure of the dilation of the blood vessels of your skin. This prevents the blood from acting as a coolant to carry away excess heat, which is then exchanged with the air in your lungs and exhaled. The symptoms of heat prostration include weakness, dizziness, vertigo, headache, nausea, blurred vision, and mild muscular cramping. Your skin will become strangely cold and you'll sweat profusely although your body temperature remains normal. This is usually a transient condition and can be easily treated if you recline in a cool environment, loosen or remove clothing, and drink copious amounts of cool water to replace sweating-caused fluid loss. Rarely does heat prostration progress to the point of circulatory collapse, which would require the immediate attention of trained medical personnel.

Heat cramp has long been experienced by workers, which is why it's also called Stoker's Cramp, Fireman's Cramp, Miner's Cramp, and Cane-Cutter's Cramp. It's almost certain that there will be a Spaceman's Cramp, too. Muscular cramps are the result of the failure of your body to replace sodium chloride lost through profuse sweating—the kind of sweating that accompanies heavy physical work. Heat cramp happens fast and it's painful. But it's transient and easily treated by relaxing in a cool place, taking salt in the form of saline water or tablets, and drinking lots of water. Heat cramps may be prevented by taking one to two grams of salt orally with water four times a day.

Hyperexia is perhaps the worst and most hazardous form of hyperthermia because it's a profound disturbance of your body's basic heat-regulating mechanism. Initially, the symptoms of hyperexia, or heat stroke, may resemble those of heat prostration combined with heat cramp *except* that your pulse rate rises to 160 or more, there's no sweating, your body temperature rises rapidly to 105°F or more, and your skin becomes hot, flushed, and dry. Hyperexia is an *emergency* situation that must be treated immediately to prevent profound shock, convulsions, cardiac failure, brain damage, and death. Hyperexia is a serious threat to life; mortality may run as high as 20 percent. As the first-aid books say, "Heroic measures must be instituted immediately." The initial task is to reduce body temperature by immersion in an ice-water bath or wrapping in a blanket soaked in cold water. The services of a doctor are mandatory because of possible changes in blood electrolyte balance. Hypothermia may occur once the hyperthermic condition is reversed, and therefore the constant attendance of trained medical personnel is required.

Effects of Low Temperatures

Exposure to cold conditions beyond the tolerance level may result in one or a combination of hypothermia injuries. The parts of your body primarily affected are the extremities, where injury may involve functional or structural disturbances or failure of small blood vessels or nerves. Your body tissue itself may

Figure 5-7. The control panel for the life-support and environmental-control system of the NASA Space Shuttle Orbiter (*Rockwell International Corporation*) (*See facing page.*)

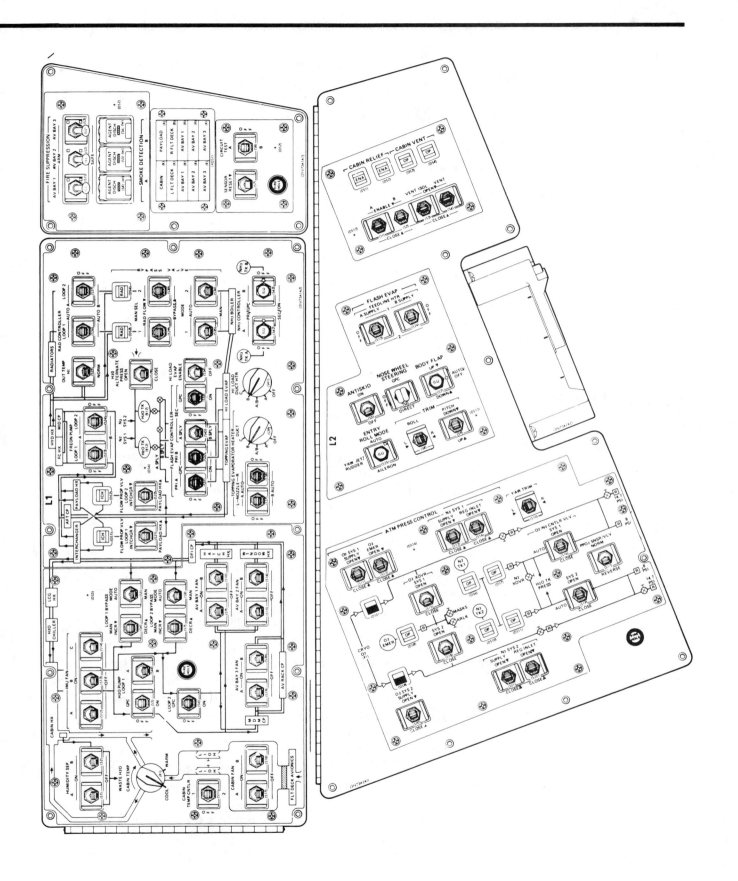

77

become frozen, causing crystallization of the water in the affected cells. Hypothermia itself may be accidental (sometimes called "exposure") or acute, which leads to a depression of your body temperature and a failure of your body's temperature-regulating mechanisms.

Exposure to high-humidity cold may result in frostnip (mild frostbite), chilblains, or trench foot; while exposure to dry cold usually causes frostbite and acute hypothermia.

Treatment of hypothermic conditions involves slowly rewarming with warm baths, hot water bottles, and heating pads. Such rewarming must be done with water temperatures no higher than 110°F to prevent burns or scalds. Tissue damaged by frostnip or frostbite should be dressed, and a tetanus booster should be administered. Full treatment should be put in the hands of a doctor.

This is particularly true of acute hypothermia which, like hyperthermia, is an *emergency* condition requiring immediate treatment by warming. During treatment and recovery, attendance by a doctor or medical personnel is required because, in addition to possible tissue damage caused by freezing, there may be problems with metabolism and electrolyte balance.

Control of the Thermal Environment

Like the environmental factors of atmospheric pressure and composition, control over the temperature and humidity environment is required for space living. In essence, you must take your familiar terrestrial environment along with you. If you don't, you face the possibility of discomfort, loss of efficiency, and the hazards of hyperthermia and hypothermia.

Your environment must have a temperature between 65°F and 90°F with a relative humidity between 30 percent and 50 percent. Outside of this limited temperature-humidity range, your performance begins to be affected and, if the temperature departs too far from the tolerable range, serious physiological problems can result which may, in the extreme, lead to death.

Proper and careful design of the life-support system of a spacecraft or space habitat will create and maintain this temperature-humidity environment, and there are many technical solutions that accomplish this. The actual system used for this purpose will depend upon the type and size of the spacecraft or habitat, when it was built, and where it is located. There are no particularly difficult problems involved in the design, construction, or operation of such systems. People started using them when the first cave fire was lit to warm this primitive living area. There are centuries of accumulated know-how and more than enough scientific data and technical knowledge to create such comfortable environments. People have been traveling into space and living in space since 1961, and the technology of space life-support systems has been evolving ever since.

Maintaining comfortable temperature-humidity environments is nothing new. People have been living safely in some extremely harsh terrestrial temperature-humidity environments for uncounted centuries. In space, however, you can no longer take these commonplace environmental factors for granted, and you'll have some control over them by virtue of proper design of the systems to maintain those required conditions. You'll no longer be completely at the mercy of the environment. As a matter of fact, snug in terrestrial homes and offices, people are no longer completely at the mercy of the forces of nature even now.

Astronaut under acceleration in a centrifuge *(NASA)*

6 Acceleration

Because of our development of transportation devices with greater speed capabilities, in the last two hundred years human beings have been subjected to increased levels of a force that has always been around us. This force is called *acceleration*. People who live in space will have to be thoroughly familiar with acceleration forces and their effects on the human body, because space transportation devices are the fastest machines ever developed.

But it isn't *speed* itself that creates the various physiological effects of acceleration. There's a considerable amount of basic truth in the quip, "It isn't speed that kills; it's the sudden stop." In fact, the saying explains acceleration neatly.

When the first fast transportation system—steam-powered railways—was put into use in the 1830s, many people believed that passengers couldn't withstand the "horrendous speed of thirty miles per hour" attained by the early railway trains. It turned out that human beings can withstand any speed if they're in an enclosed compartment protected from the blast of air rushing past.

Compared to the stagecoach or canal barge, the railway train was able to change speed rapidly, and this was the aspect of speedy travel that had a demonstrable effect on people. Jolting starts and sudden stops—to say nothing of the sudden change in speeds that are a part of any collision—tossed passengers around with forces much higher than those created by being thrown from a horse.

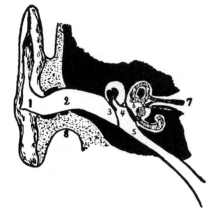

Figure 6-1. The semicircular canals of the human ear will detect *rotational* motion or acceleration. Key: (1) ear or *pinna*, (2) external auditory canal, (3) eardrum, (4) middle ear containing the ossicle bones, (5) Eustacian tube to throat, (6) inner ear and semicircular canals, (7) auditory nerve to brain, (8) parotid salivary gland
(© 1985 Federal Aviation Administration)

In short, we have no physiological mechanism, no sense organ, that can detect speed per se, but we can and do sense and react to *change of speed* or *acceleration*.

What Is Acceleration?

Acceleration is defined as a change of velocity per unit time.

Although many people think of speed and velocity as synonyms, they are not: velocity is a function of both speed *and* direction. And velocity can be changed by altering either speed *or* direction.

When any object changes velocity, it accelerates. Everything inside it or attached to it and traveling along with it is subjected to the same acceleration. You sense this acceleration as a force that is basically caused by the inertial effects of mass—i.e., anything that's in motion wants to continue moving at that speed in a straight line, and a change in speed or direction therefore produces a force opposing the change.

Sensing Acceleration

If you make an automobile increase speed suddenly by stepping on the accelerator, you'll be pushed back into the seat cushions. If you apply the brakes to slow the car quickly, you'll be thrown forward. When you make the car turn quickly, you change its direction and feel a sideward force. These are mild forms of acceleration that don't greatly affect you although you can sense them. They're low accelerations and don't last very long. They are well within your acceleration tolerances.

But when velocity is changed rapidly as in a crash or impact, you are subjected to high acceleration forces over a short period of time, and these forces do create physical hazards.

In a like manner, there are definite physical hazards that accompany high accelerations over a long period of time. These are the sorts of accelerations you'll experience in a spacecraft.

A spacecraft is just like an automobile in this respect: It's capable of accelerating. It may be capable of traveling at a speed of several miles per second, but it takes time to reach this speed. Time and the change of velocity that occurs during that period of time are the factors that define its acceleration.

When you're in a spacecraft and change its speed, you'll feel the acceleration forces just as you do in an automobile. But the acceleration forces of space travel can be five to ten times as high as those of automobile travel and are of the same magnitude as the acceleration forces encountered in aerobatic airplanes.

Anyone traveling into or living in space must have a basic understanding about acceleration. On Earth, you can "take it or leave it," so to speak, because it doesn't seem to be an imporant factor in your everyday life (although indeed it may if you ride in any transportation device). However, living on the surface of Planet Earth has given you a warped or distorted mental image of the way the rest of the universe works. You're used to the unique characteristics of planetary living, which include the constant presence of a gravitational pull and the ability to keep such things as acceleration forces within tolerable ranges of magnitude and duration—unless you fall or have an automobile accident.

Understanding must be followed by measurement because, according to Lord Kelvin's admonition quoted earlier, when you measure, you know and can predict.

Acceleration and Gravity

The effects of gravity are similar to those of acceleration. In fact, without special measuring instruments, you can't tell the difference between acceleration and gravitation. This is the "principle of equivalence" developed by Albert Einstein and one of his concepts that's easily understood by almost anyone.

If you were in an enclosed elevator with no way to receive information on what was happening outside, you couldn't tell whether gravity was holding your feet to the elevator floor or if the elevator were moving upward with an acceleration that produced a force equal to that produced by gravity.

A gravitational field acts as though it were accelerating mass. Nobody really knows *why* a gravitational field does this or what gravity really is yet, although there are a lot of hypotheses and theories about it. The gravitational field of Earth always pulls you toward the center of the planet. If the ground wasn't under your feet, you'd fall toward the center of the Earth. But since the surface of the Earth prevents you from falling, it exerts a force on you that you can't sense as being any different from acceleration force.

Measuring Gravity

For convenience, engineers usually use the term "g" or "gee" for measurement of acceleration because this compares it to the pull of Earth's gravity.

Standing on the surface of the Earth, all of us experience an acceleration force of 1 g. This is equivalent to being accelerated with a velocity change of

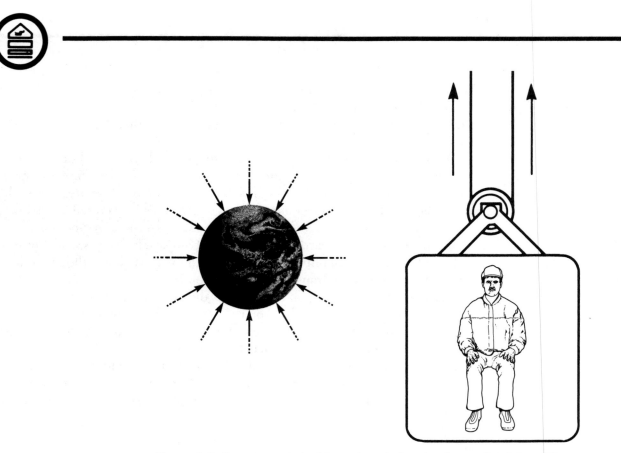

Figure 6-2. If a person is inside a closed elevator, he can't tell the difference between the acceleration of gravity or the acceleration caused by change of velocity. *(Art by Sternbach)*

32.17405 feet per second every second, which is rounded off and written as 32.17 ft/sec/sec or 32.17 ft/sec². This gravitational acceleration isn't the same everywhere. It's different at different places on Earth and decreases as altitude above the Earth's surface increases. It's only when you have to compute highly accurate spacecraft flight paths near the Earth that the actual gravitational acceleration must be used in calculations. Therefore, although the gravity field of the Earth varies, the Earth's "standard gravitational acceleration" of 32.17 ft/sec² is used for most rough "back of the envelope" calculations.

Two g's would be an acceleration equal to twice that of gravity, or an acceleration of 63.34 ft/sec².

This can be illustrated if you get on the bathroom scale and the scale reports you weigh 150 pounds. It means you're being pulled toward the center of the Earth, exerting a force of 150 pounds on the bathroom scale because of gravity. If the Earth's gravity were twice as strong—i.e., if it were 2 g's—the bathroom scale would show you to weigh 300 pounds.

Now imagine that you take the scale and get aboard an elevator. At rest in the Earth's gravity field, the scale would show you to weigh 150 pounds. Now start the elevator upward so that it gains speed in the amount of 32.17 feet per second every second. You'd be subjected to 2 g's of acceleration—1 g from the Earth's gravity field and 1 g from the acceleration of the elevator. You'd feel and act as if you weighed the 300 pounds indicated by the scales.

Acceleration Effects

Automobiles rarely achieve accelerations of 1 g, but airplanes can achieve accelerations of 10 g's or more while turning. Rocket vehicles can achieve accelerations as high as 100 g's.

Electronic instruments and other mechanical devices can be built to withstand the crushing force of 100 g's, which makes them seem to weigh a hundred times normal; but people can't.

At 100 g's acceleration, you'd weigh 100 times what you do on the Earth's surface. Bones and muscles simply cannot withstand those forces, and you'd be crushed by your own mass.

The Increasing Need for Data

Most of the research on human tolerance to acceleration was carried out in the period between 1945 and 1965. Military pilots during World War II reported that the accelerations attained during combat maneuvers were causing them to lose consciousness. Pilots discovered that if the acceleration force pulled them down into the seat, they'd "black out." This was caused by the acceleration force pulling the blood from their heads and causing their brains to become temporarily starved for oxygen. This resulted in loss of vision and, if the acceleration continued long enough or if it was great enough, unconsciousness. Acceleration in the opposite direction created when performing an outside loop would cause blood to rush to their heads and make them "red out," see red, instead. Accelerations that pushed them back into their seats ("eyeballs in" acceleration) was easier to withstand than the opposite ("eyeballs out") because their seats supported their bodies during the acceleration. But how many g's could a pilot withstand before his critical mental and physical abilities were impaired to the point where he couldn't fly and fight? Military requirements to design and build faster and more maneuverable combat airplanes therefore required more data on acceleration tolerances to prevent the airplanes from killing their pilots.

Engineers who conducted early pre-Sputnik studies of space travel followed by the design of the first manned space capsules needed information on how much acceleration a person could withstand. Since the early manned space boosters were converted military ballistic missiles, the g-forces on any space capsule using these rockets could be as much as 9 g's. Could a human being withstand such acceleration for periods of several minutes? Would the astronaut be able to exercise control over the capsule during high acceleration periods? What was the best position for the astronaut to be in with respect to the direction of the acceleration? How much acceleration could the astronaut withstand during reentry and landing? The answers to these questions determined many design factors of the Mercury and Gemini space capsules, including such things as the size of the parachutes required to keep landing accelerations and shocks to an acceptable level.

(Whereas the first U.S. space capsules landed in the ocean to keep the

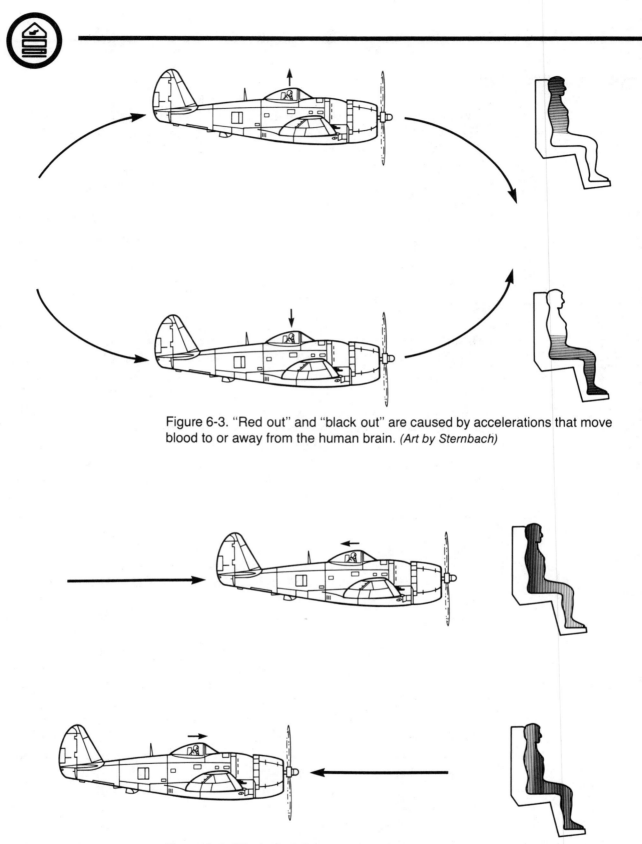

Figure 6-3. "Red out" and "black out" are caused by accelerations that move blood to or away from the human brain. *(Art by Sternbach)*

Figure 6-4. "Eyeballs-in" and "eyeballs-out" accelerations are graphic terms describing the effects of fore-and-aft accelerations on a human being. *(Art by Sternbach)* landing forces within human tolerance, the Soviet Vostok touched down on land so hard that the cosmonaut was required to eject from the descending Vostok and come down on a separate parachute.)

As body-contact sports such as football became more rugged and stressful, there were indications that impacts were severely injuring athletes in spite of protective padding. What were the human tolerances to short-duration acceleration with a very high rate of acceleration onset or jolt? Data was obviously needed in order to completely redesign much existing athletic equipment and develop new equipment as well.

Finally, the increasing number of automobile accidents that caused serious injuries and death led the federal government to demand safer automobiles that wouldn't kill their occupants by collapsing and crushing them or by throwing them violently against sharp or impaling interior elements such as knobs, steering columns, or decorations. And the United States Air Force wanted to know how much g-force a pilot could take before he'd lose control of the plane. In addition, the Air Force discovered it was losing more pilots in automobile accidents than in plane crashes.

These demands for information required a deeper understanding of human tolerance to accelerations and to "jolt" or a change of acceleration.

Getting Acceleration Data

Before astronauts and cosmonauts began riding rockets into space, a great deal of research was carried out on Earth to determine the limits of human tolerance to acceleration. Researchers used centrifuges that spun volunteer human subjects around in capsules at the end of long mechanical arms, thus creating high and sustained accelerations by centrifugal force.

Figure 6-5. Huge whirling centrifuges can be used to test human tolerances to long-term effects of moderately high accelerations. *(North American Aviation, Inc.)*

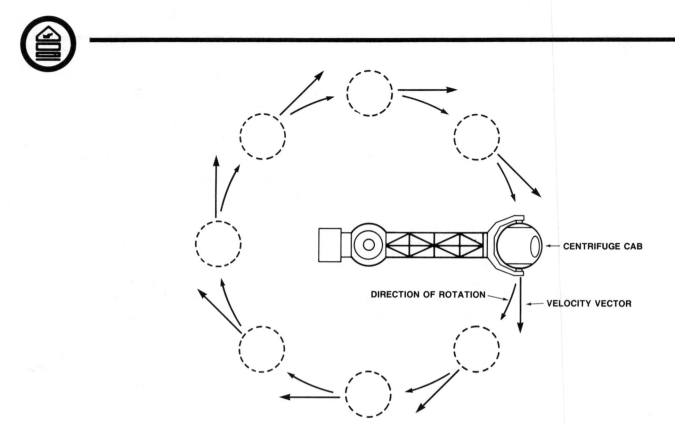

Figure 6-6. A centrifuge produces a true acceleration, which is a change in the magnitude or direction of velocity, by constantly changing the direction of the velocity vector as shown. *(Art by Sternbach)*

Centrifugal force is really an acceleration force caused by a constantly changing direction at constant speed as shown in Figure 6–6. The person in the centrifuge cab has mass that possesses inertia, which in turn wants to keep everything going in a direction tangent to the circle of rotation. However, the centrifuge arm keeps pulling the cab and human subject inward, constantly changing the direction of the velocity component.

Researchers also used horizontal accelerators. Some of these were rail-mounted devices propelled by rockets horizontally along a railwaylike track on the ground. They were stopped by scoops that extended into water troughs between the rails of the track. Other horizontal accelerators were moved rapidly by air-driven pistons to provide short durations of acceleration, whose data were useful in crash studies.

Human Acceleration-Tolerance Factors

It was generally agreed among experts in 1940 that 10 g's was about the limit of human tolerance to sustained acceleration. There was no scientific or experimental basis for this. It was an educated guess mostly based upon reports from pilots who'd been subjected to high accelerations in high-performance fighter or "pursuit" airplanes. During World War II, combat maneuvers created forces

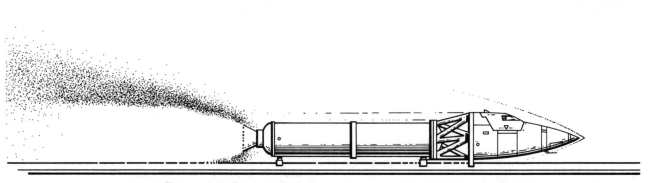

Figure 6-7. Rocket sleds have been used to test the effects of short-term accelerations on human beings. *(Art by Sternbach)*

as high as 9 to 10 g's on pilots. This was compounded when high-speed jet airplanes first appeared.

Dr. John Paul Stapp, an Air Force flight surgeon, pioneered acceleration, jolt, and crash studies and became famous because he wouldn't let subordinates and experimental subjects participate in any tests on rocket sleds unless he'd personally gone first. Using centrifuges and rocket sleds, Stapp and other Air Force human factors researchers discovered that human tolerance to acceleration involves a combination of four main factors:

1. the *direction* of the acceleration force with respect to the human body

2. the *intensity* of the acceleration force

3. the *duration* of the acceleration force

4. the *rate of onset* or the suddenness with which the acceleration is applied to the human body

These four factors are intimately interrelated.

Constant Acceleration-Tolerance Limits

Insofar as long-term tolerance to acceleration is concerned, the primary factors are the direction and intensity of the force.

The positions of greatest and least sustained acceleration tolerance are shown in Figure 6–8.

You have the greatest tolerance level when you're immersed in water lying on your back with your torso and head inclined upward at an angle of 35°. Surprisingly, you have less acceleration tolerance when you're lying flat on your back.

Without water immersion, which could add weight and complexity to any travel in a spacecraft, your best position for maximum tolerance to sustained acceleration is a reclining sitting position, upper legs or thighs at right angles to the acceleration and your head and torso inclined forward at a slight angle against the direction of acceleration. This position is the one that's been

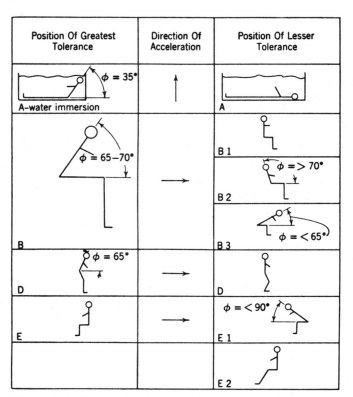

Position Of Greatest Tolerance	Direction Of Acceleration	Position Of Lesser Tolerance
φ = 35° A—water immersion	↑	A
φ = 65–70° B	→	B 1 / φ => 70° B 2 / φ = <65° B 3
φ = 65° D	→	D
E	→	φ = <90° E 1
		E 2

Figure 6-8. The position of a human being with respect to the acceleration has a great deal to do with his acceleration tolerance. *(U.S. Air Force)*

universally used in all American and Soviet manned spacecraft to date, including the NASA Space Shuttle Orbiter. However, in some recent high-performance fighter aircraft such as the F-16 "Fighting Falcon," the pilot reclines at a greater angle with his legs more in the position shown in Figure 6–9.

Figure 6–10 shows the effects of position on time tolerance expressed in terms of "g-minutes" (one g-minute equals 1 g applied for 1 minute). Figure 6–10 was compiled for use in the design of aircraft and spacecraft for crew seating in various orientations—backward or "eyeballs out," forward or "eyeballs in," foot-to-head, and with the person immersed in water.

Standing erect with acceleration from head to foot as you're used to it standing on Earth, you can withstand 2 g's almost indefinitely. It's like carrying someone else around on your shoulders—difficult, but bearable. You can do it, but you soon become fatigued unless you've conditioned yourself to it by practice and regular exercise. But it's impossible to remain standing at an acceleration of 3 g's; walking and climbing become impossible, and crawling is difficult at this acceleration; and the soft tissues of the body begin to sag. At 4 g's, crawling becomes impossible and even movement itself is possible only with great effort; the only position you can maintain is the sitting position. At 5 g's, only slight movements of your arms and head are possible. Visual symptoms such as "gray-out" begin to appear at 2.5 g's, and at 7 g's, blackout usually occurs. At 8.5 g's, your heart can't force enough blood up to your brain, and

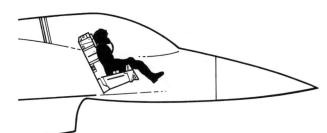

Figure 6-9. In both the Space Shuttle Orbiter and a high-performance fighter plane such as the F-16, seats are designed and positioned so that the high accelerations can be easily tolerated by the human occupant. *(Art by Sternbach)*

unconsciousness sets in. An acceleration of from 18 to 23 g's causes structural damage, especially to your spinal column.

With the acceleration in the opposite direction—foot to head or "head down"—the tolerance limits are *much* less.

However, when you take acceleration *across* your body rather than from head to foot, you can tolerate constant acceleration for a much longer time at a higher level. Human subjects have withstood constant accelerations of up to 17 g's in this supine position with no loss of vision or consciousness. Auditory function remains intact up to 14 g's. You can tolerate up to 30 g's in this position without structural damage. Although head and arm movements become impossible at 6 g's, you can move your wrists, hands, and fingers up to 12 g's.

On the other hand, with the possible exception of military spacecraft, it is unlikely that you'll encounter sustained accelerations above 3 g's because it's unnecessary to exceed this for nearly all nonmilitary space missions. The NASA Space Shuttle, the first space transportation system specifically designed from the beginning for people, limits its sustained acceleration to 3 g's by throttling its rocket engines and by proper control of its entry and landing path.

Constant or relatively long-term accelerations occur over periods ranging from several minutes to an hour. As the duration of high acceleration increases, human tolerance decreases. It might be possible for a person to withstand 4 to 6 g's for several days, but it would be difficult to carry out normal biological functions while pinned down to a couch by this level of g-force. There is also little or no indication of potential physiological effects of long-term exposure to high levels of acceleration.

However, as the duration of acceleration grows shorter, human tolerance becomes greater, but another factor enters the picture: the rate of onset of the acceleration or the degree of jolt.

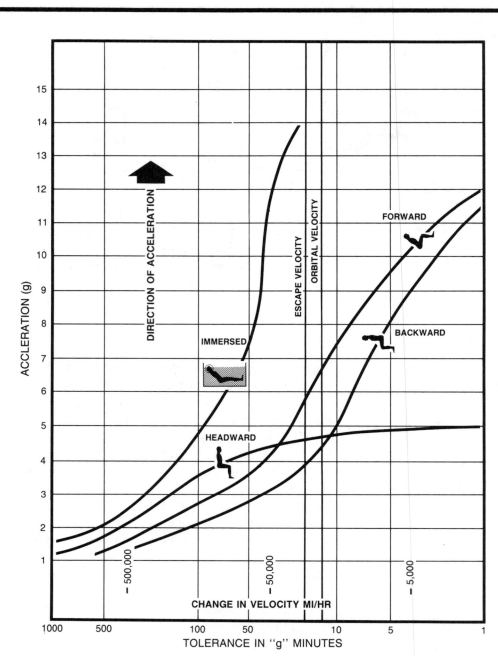

Figure 6-10. Human tolerance to acceleration as a function of acceleration, direction, and time *(Art by Sternbach)*

Short-Duration Acceleration-Tolerance Limits

If one second is required to go from 1 g to 10 g's acceleration, the rate of onset is 9 g's per second. This is a mild jolt. Applying 100 g's at a rate of 1,500 g's per second for a hundredth of a second is equivalent to banging your fist on a table, but that's only a localized rate-of-onset effect that can be easily tolerated. However, if 100 g's at 1,500 g's per second is applied to the entire body, a person can survive only if properly supported and/or protected. As the rate of onset approaches 3,000 g's per second, extreme crash conditions are involved and structural damage caused by the physical shock can become lethal—e.g., the rupture of major blood vessels.

Aerospace medical researchers have learned that high impacts and rates of onset are encountered by people even in ordinary athletics. The author helped instrument two healthy college football players with devices to measure the acceleration and rate of onset they experienced when blocking and tackling one another. They experienced 70 g accelerations at a rate of onset of 2,000 g's per second for as long as 0.2 seconds. This data, taken in 1960, helped establish the United States Air Force criteria for ejection seats and escape capsules used in military aircraft.

Other data on human tolerances to acceleration and jolt have come from strange places. To get information on the strength of the human neck and to determine whether or not head restraints would be required in aircraft ejection seats, aeromedical researchers studied old books on hanging. They then wired an anthropomorphic dummy with instruments and hung it from the gallows for the benefit of the recording instruments. The weak-hearted members of that group declined to carry the investigation any further when they dropped the dummy ten feet and pulled its head off. At that point, they'd learned the limiting strength of the human neck.

The absolute top of the acceleration tolerance spectrum is an oddity. A paratrooper whose parachute had failed landed on his back in the soft dirt of a recently plowed field and walked away with only a sprained wrist. He'd survived an estimated 200 g's for 0.015 seconds.

Protection against Acceleration and Jolt

As a result of these studies and tests, it's been learned that when you're subjected to acceleration and jolt, you act like a floppy bag of gelatin that gives under acceleration forces and bounces. Protecting yourself against acceleration and jolt involves proper cushioning and support to distribute the forces over as much of your body as possible. You've also got to keep yourself from bouncing around inside the vehicle.

Seat belts in automobiles and aircraft *do* make it much more likely that you'll survive an impact. Add the cross-chest strap, and your chances of crash survival improve even more. Although it's been impossible to get all automobile drivers and passengers to buckle up, Federal Aviation Administration regulations *require* the use of seat belts and shoulder harnesses in aircraft, and the use of such restraints is taught to student pilots from the first time they get into an airplane.

The best restraint harness has been in use for decades in military and aerobatic aircraft: a lap belt, a strap going up over each shoulder, and a crotch strap to keep you from sliding out under the harness. With an inertia reel on the shoulder harness to permit you to lean forward, and with a quick-release latch that holds the harness together over your lap to permit you to quickly

93

Astronaut Leroy Cooper in typical restraint harness for protecting against acceleration. *(NASA)*

release the harness by hitting a single large latch with your fist, this harness will properly restrain you against very high acceleration forces, especially in the eyeballs-out direction that's extremely critical in most crash situations. The use of this type of strap restraint system will continue in spacecraft because, unlike many other restraint garments and systems that have been proposed in the past, it's economical and fits nearly everyone.

Modern cabin and vehicle interior design has eliminated most of the sharp projections and edges that can cause injury if you strike them during the rapid acceleration or jolt of a crash. However, deeply padded surfaces often are not the best for catching you without injury because they can cause your body to rebound. In some cases, the best way to catch and cushion you is on a sheet of steel that will deform and stretch without rebounding, or into a plastic foam or honeycomb structure that will collapse and absorb the energy of impact.

It has continually amazed researchers to discover that you are more rugged than the devices you travel in. Modern studies of human acceleration and jolt tolerances have shown that, in nearly all tests, properly restrained and supported people would have withstood the crash forces only to be killed by the vehicle structure collapsing and crushing them.

This is one area where aerospace research has paid off in our everyday lives on Earth.

But anyone who lives and works in space will be very conscious of acceleration at all times not only because of the accelerations experienced in space travel but because the space environment itself has a unique characteristic that can't be duplicated for more than a few seconds on Earth: weightlessness.

The NASA Skylab gave people their first opportunity to move about in zero-g. *(NASA)*

7

Weightlessness

Weightlessness, or "zero-g," is a characteristic of space that cannot be experienced by people on the surface of the Earth except for brief exposures of less than a minute.

Weightlessness is a consequence of the manner in which spaceships travel in the Earth-Moon system and especially over the longer trajectories of the solar system, and also of the peculiarities of orbital mechanics. For most space travel and in most space facilities, the force of gravity will appear to be absent.

The effects of weightlessness on your body and even your mind are more profound than the effects of acceleration because in space you'll be living and working in the weightless condition most of the time unless you're in a habitat that has part of it rotating to produce the pseudogravity of centrifugal force.

The Reason for Weightlessness

Some people assume that weightlessness exists because the spacecraft or habitat is beyond the pull of gravity or out of a gravitational field. Nothing could be further from the truth. Weightlessness is caused by a definite type of motion within a gravitational field.

Like the Earth's atmosphere, the Earth's gravity field slowly decreases in strength or ability to accelerate an object with increasing distance from the Earth until it finally becomes weaker than the gravity field of the Sun. In the

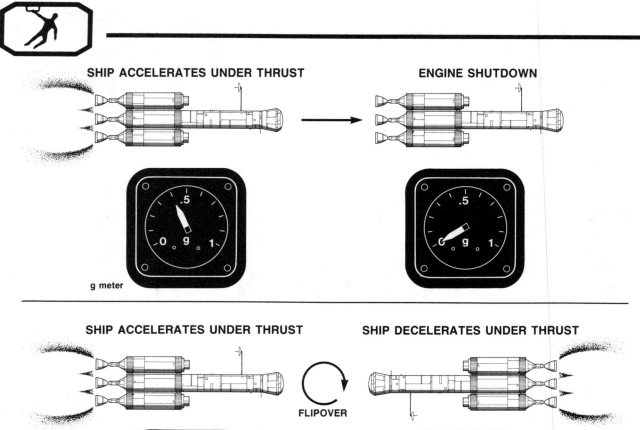

SHIP ACCELERATES UNDER THRUST

ENGINE SHUTDOWN

g meter

SHIP ACCELERATES UNDER THRUST

SHIP DECELERATES UNDER THRUST

FLIPOVER

Figure 7-1. In space travel, a person will feel "weight" only when the spaceship's rocket motors are thrusting. For classical flights, there will be acceleration only at the start and end of the flight with no acceleration in between. Future spaceships make "constant boost" flights with a flip-over at midpoint, creating acceleration forces during the entire trip. *(Art by Sternbach)*

Earth-Moon system, this means that nothing is ever beyond the influence of Earth's gravity. And in the solar system, nothing is ever beyond the pull of the sun's gravity.

Why are you and your habitat in a weightless condition if you're never beyond the influence of a gravitational field?

The rocket-powered spacecraft as it currently exists and as we can project its design into the foreseeable future will be an "impulse" type vehicle. It achieves the necessary velocity to shape its path through space to its destination by the application of propulsive force that accelerates it during the first minutes of flight. Once having attained this velocity, the rocket propulsion system is then shut down. The spacecraft then coasts through space to its destination. There a final application of rocket impulse is made to match velocities with the destination, whether this be a planet, satellite, space station, or another spacecraft.

98

This is the most efficient way to travel between any two points in the Earth-Moon system or in the solar system using the rocket engine as we know it today. It requires the least amount of rocket propellant and therefore energy.

The only difference between travel in the Earth-Moon system and the solar system is the time involved to make the trip, which in turn is a function of the velocity to which the ship can accelerate. But that's a subject for a book about propulsion and celestial mechanics, and there have been many books of that sort published.

At some time in the indeterminate future, other types of space-propulsion systems may become available to permit "constant boost" travel—i.e., constant acceleration at, say, 0.1 g out to the midpoint of the flight where the craft begins to decelerate at 0.1 g to its destination. In constant-boost flight, there's no weightlessness or zero-g. And travel times become fantastically short, even at constant accelerations of a fraction of a g. A flight from the Earth to the Moon at a constant boost of 0.1 g would require only 10 hours, 51 minutes, and the maximum velocity at "turnover" would be 12.1 miles per second or 43,524 miles per hour. Constant-boost flight really cuts the solar system down to manageable size in terms of trip times because a constant-boost flight to Mars under the worst possible conditions with respect to planetary positions would require at the most a total of 10 days under a constant boost of only 0.1 g. At a constant 1-g boost, the Mars trip time comes down to 4.52 days. There's one overwhelming problem with forecasting a constant-boost propulsion system: nobody yet has the foggiest notion of how to build one. There are theories, and there are a few gadgets that may or may not be "proof of principle" devices. But no one has yet demonstrated a "space drive." So we'll have to stick with primitive old rockets for the foreseeable future.

Therefore, a spaceship that's coasting out to its destination with its engines shut down and a space habitat in orbit both fall freely in a gravity field.

In the case of a spaceship going away from Earth in coasting flight, it falls *up*. Approaching Earth, it falls *down*. This is not unusual behavior. Many objects fall both up and down.

To gain a better understanding of how something can be weightless while it's falling, throw a hollow rubber ball such as a tennis ball, paddle-tennis ball, or handball into the air. The motion of your arm and hand imparts a velocity—speed and direction—to the ball that leaves your hand after being accelerated over a distance of about 5 feet. The ball will, of course, travel a much greater distance than the 5 feet through which it was accelerated, depending upon the speed and direction you imparted to it by throwing it.

The ball begins to fall from the moment it leaves your hand because it's no longer being accelerated by your arm and hand motion. First the ball falls up, then it falls down. If the drag of the Earth's atmosphere is neglected, the only force acting upon the ball once it leaves your hand is the Earth's gravity, which draws toward the center of the Earth the ball and anything in it or on it.

Now imagine that you're inside the ball and traveling with it. Anything attached to the ball or inside the ball travels with it and falls freely with it, including you. There's no force that you would or could perceive as "gravity" acting between you, the ball, and anything on or inside the ball.

The "throwing arm" of any orbiting satellite is the launch vehicle that

accelerates it from the surface of the Earth to orbital altitude where the satellite is given the proper velocity—speed and direction. Once placed in orbit, the satellite then falls constantly toward the center of the Earth just like the ball in the example. But the Earth's surface constantly curves away from it so that the satellite never reaches the Earth's surface. When we talk about satellites in Earth orbit, the simple fact that the Earth is a spherical must be taken into account.

Newton's Cannonball Analogy

The best analogy of satellite motion and weightlessness was given by the man who originally worked out the mathematics of the law of gravity, Sir Isaac Newton. In his pioneering work on mechanics, *Philosophiae Naturalis Principia Mathematica* (known as the *Principia*), published in 1687 in England—with the financial assistance of Sir Edmund Halley, for whom the comet is named—Newton imagined a planet like the Earth with a high mountain on its surface whose summit extends above the top of the planetary atmosphere.

Put a cannon at the summit as shown in Figure 7–2 and fire a cannonball horizontally. The cannonball will start to fall the instant it leaves the gun's

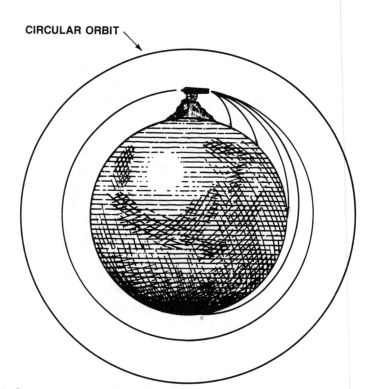

Figure 7-2. Sir Isaac Newton's explanation of satellite orbits and velocity using a cannon to launch shells at increasing velocity until finally one falls all the way around the world *(Art by Sternbach)*

muzzle and will strike the surface of the planet at some given distance from the mountain.

Reload the cannon with more gunpowder, and fire the cannonball again with even greater muzzle velocity; the cannonball will go much farther before it finally reaches the ground.

There is a condition, Newton pointed out, where a large gunpowder charge in the cannon would give the cannonball enough muzzle velocity so that it would go all the way around the planet and hit the cannon from behind. At that point, you've made a shot heard 'round the world and, if you remove the cannon in time, you've got a satellite that will continue to fall around the planet forever, just like a little moon.

This phenomenon is often explained by saying that the centrifugal force generated by the velocity of the satellite exactly balances the gravitational force. This can be considered to be true. But the more general case of the constantly falling cannonball and the constantly curving surface of the planetary sphere of Newton's analogy will hold true and provide a better explanation of the phenomenon for nearly all cases of orbiting bodies.

The Sensation of Weightlessness

How does this produce the sensation of weightlessness?

When you sit in a chair on Earth, you feel the weight of your body against the chair. Actually, you're feeling the Earth's gravitational field attempting to accelerate your body toward the center of the Earth. If the chair, the floor, the building foundation, and the ground were not under you, you'd indeed fall unrestricted toward the center of the Earth. Since the acceleration of Earth's gravity field is 32.17 feet per second per second, you'd fall at a speed of 32.17 feet per second at the end of the first second, 64.34 feet per second at the end of the second second, and so forth. At the end of the first second, you'd fall 16.09 feet; at the end of the second, 64.34 feet, and so forth. You'd have the sensation of falling, and you'd brace yourself for the sudden stop that your experience tells you *always* happens sooner or later.

The weight you feel while sitting in the chair is caused by the resisting force of the chair against the acceleration of the gravity field. Remove that resistance, and you fall freely in a weightless condition.

The Falling Elevator Analogy

To reinforce this explanation, imagine yourself and the chair in the closed elevator car suggested in the previous chapter to explain Einstein's principle of equivalence between acceleration force and gravitation force. If somebody cuts the elevator cables and the elevator car begins to fall freely down the shaft, you'll fall with the chair and with the elevator. And you'll feel no force between your bottom and the chair. You'd be falling along with the chair and the elevator.

Our Earthly Experiences

Weightlessness is a condition you've never encountered on the Earth's surface except for very short periods of time. You live with gravity day and night, and it's an ordinary and normal thing to you. You expect it to be there, and you really don't think about it. When it isn't there, or when the sensation of a force between your body and a support disappears, it means you're falling. Alarm bells go off in your brain because you've learned from long experience that all falls *always* end in sudden and painful stops. The longer the falling sensation lasts, the worse you know the bump is going to be.

You're not alone in your human adaptation to Earth's gravity. All life on this planet evolved in Earth's gravity field and has adapted to it. Plants send their roots downward and stalks upward in response to the tug of gravity, a phenomenon called *geotropism*. Although sea animals gain most of their bodily support from buoyancy, land animals have skeletons and muscles to support their bodies against the force of gravity. Some animals have better gravity adaptation than others. Strangely, most large mammals are poorly equipped in this regard. Dogs, horses, and people have back problems, hip problems, and other maladies that are the result of structures ill-equipped to support them in a gravity field.

Occasionally, you're subjected to a brief period of weightlessness when you're falling freely—the initial descent of a fast elevator, jumping off a diving board, bouncing on a trampoline, or skydiving with a parachute. But these periods of weightlessness last at most only a second or so.

In space, weightlessness is normal and the acceleration of gravity isn't.

The Effects of Weightlessness

Before the Space Age began in 1957, we didn't know much about the effects of weightlessness upon your body, how you'd react to constant falling, or whether you could live without a gravity force acting upon your body. As the years have gone by, and as more people have flown and lived in space for increasing periods of time, we've gotten better answers and more data. Some of the answers have been positive, and some haven't. But, in general, on the basis of the studies and measurements made on all the people who've traveled into space thus far, here's what we know:

For short periods of time up to about six months, you won't have any problem living in weightlessness. Some people, however—and you may be one of them—require several hours or several days to adapt to the weightless condition. The latest data shows that you have a 50 percent chance of becoming motion sick during the first day of weightless living. You'll probably get over it within a few hours. If you can't adapt, you'll have to come home to Earth.

You'll have no trouble carrying out most normal living functions in zero-g. You don't require the presence of gravity for breathing. You can swallow both food and water without gravity. Not only have astronauts and cosmonauts done these things perfectly well, but you can even do them *against* the force of gravity.

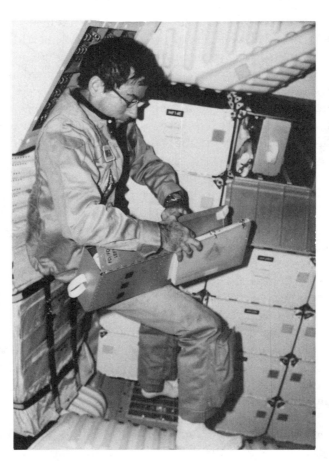

Living and working in weightlessness can pose new problems as well as solving others. John Young floats in the mid-deck of the Space Shuttle Orbiter "Columbia" with personal items and food containers floating all around him. *(NASA)*

Because there's no up or down in weightlessness, things such as medical checkups become easier to do. *(NASA)*

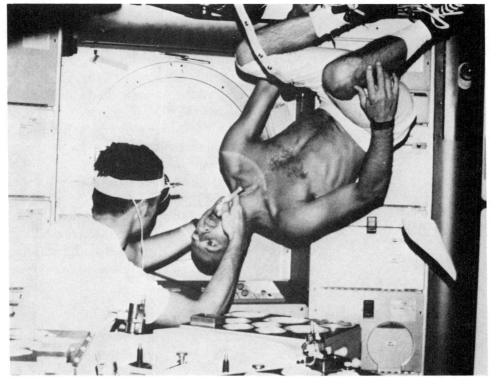

You can prove to yourself that gravity has little effect on such things. Stand on your head; you can still breathe, and you can still eat and swallow a cracker.

Elimination of human waste also presents no serious problem provided the "waste-management system" is properly designed to receive and contain such wastes. Early methods of waste management were rather primitive because people weren't going to remain in space for more than a few weeks. Zero-g toilets have improved constantly since the first one was used in Skylab. The Space Shuttle Orbiter toilet is probably going to set the standard for decades to come.

Your muscle coordination is somewhat affected by zero-g. Your hand-eye coordination appears to suffer the most if the visual component is absent. During the *Skylab* 3 mission in mid-1973, American astronaut Owen Garriott reported that crew members found it almost impossible in the darkened sleeping quarters to reach out and touch the light switch located less than two feet away. "The result was not just a near miss," Garriott reported. "We found that our hands might first encounter a locker as much as forty-five degrees away from the correct direction. Although I tried to 'practice' this move on a number of occasions, I still could not do it well after two months."

However, if you can see what you're reaching for or handling, you'll have absolutely no problem with eye-hand muscle coordination in weightlessness. Both American and Soviet space crews have worked successfully in space for months at a time and have reported no problems with hand-eye coordination if they could see what they were doing.

Furthermore, astronauts have had no trouble maintaining their orientation regarding up and down. According to former astronauts Gerald Carr and Ronald Evans, with whom I've discussed some of the problems of weightless living, when you're weightless in space, "up" is the direction your head is pointed toward. "Down" is toward your feet.

American astronauts continually report that it's easy to move around inside a spacecraft or habitat in weightlessness. In fact, you'll find it's much easier to move around a space habitat than it is to get around the simulated facility on the ground in one g. If you've ever watched a diver, a gymnast, or an acrobat, you'll realize that you're quite capable of moving and controlling your body in brief periods of weightlessness once you get a little practice. The TV pictures and movies from both Skylab and Shuttle show that people quickly adapt to getting around in weightlessness, quickly invent new methods of locomotion, and even develop weightless games and "astrobatics" or acrobatics. Not only does it look like a lot of fun, but the astronauts report that it's one of the best recreations they've discovered in space. It will take very little time for you to reprogram your reactions and overcome your instinctual falling reactions; apparently, the relief experienced in discovering that the constant falling won't result in a sudden stop is partly evidenced in such joyous fun.

Adapting to Weightlessness

However, from the very start of human travel in space, some people have adapted immediately to weightlessness while others have required up to several days. The ingrained, instinctual fear of falling is very deep-seated in all human beings because it's a survival trait. In addition, your semicircular canals or otoliths, the balance organs in your middle ear, don't know you're in weightlessness and may send conflicting and confusing signals to your brain when you try to move around, producing a situation called an "autonomic storm."

Some of the most highly motivated astronauts haven't been able to overcome these fears or the confusing signals of an autonomic storm. As a result, they've suffered from motion sickness that includes nausea and disorientation. Usually this weightlessness-adaptation syndrome disappears in a matter of minutes or, at the most, a few days.

Space motion sickness was first reported by the fourth man in space, the Soviet cosmonaut German Stepanovich Titov, who was launched in *Vostok 2* on August 6, 1961. Titov experienced an initial nausea and disorientation, but was able to adapt and adjust to weightlessness within a matter of hours.

Nobody who's gone into space as of this writing has been unable to adapt to weightlessness within a day or so although some people have had a very difficult initial few hours.

But, as mentioned above, those people were highly motivated. They worked hard, sometimes for years, to get into space. Most of them were experienced test pilots; they had the "right stuff"; and there was a certain "macho" self-image that wouldn't permit them to be sick; so they fought it down and won. It's quite probable that this very strong motivational drive enabled them to overcome and conquer their initial reactions to weightlessness. It remains to be seen what will happen when thousands of people such as you first go into space. Some of them may not be as strongly driven.

EVA Disorientation

Some American astronauts, however, haven't been able to totally adapt to the weightless condition outside a spacecraft in "extravehicular activity" or EVA. They've suffered from a phobia allied to the fear of falling: acrophobia, or the fear of high places, which has its roots in the fear of falling. Unfortunately, to protect the "right stuff" and heroic image of some of these astronauts, their phobias weren't widely reported; but they can be learned of if one digs deep enough into the actual mission reports and debriefings. As a result of a phobia, you may not be able to perform useful work outside an enclosed spacecraft or habitat, and you may not know whether or not you can until you get out on the end of an umbilical tether on an EVA, although some tests may be developed that would spot this ahead of time.

Vertigo

Some disorientation may be due to vertigo. Actually, disorient literally means "to turn from the east," or, in the vernacular, "Which way is up?" Vertigo is a severe form of disorientation, a state of temporary spatial confusion resulting

105

from your brain's misinterpretation of conflicting information received from your sense organs, primarily your eyes. When visual information doesn't match with other information coming in from your body's balance senses—the semicircular canals of the inner ear plus the kinesthetics of touch and muscle resistance—your brain doesn't know which to believe or how to sort out the conflicting information.

It's difficult to describe what vertigo is like because it amounts to almost total sensory confusion. It's possible to experience vertigo in ordinary amusement-park rides on Earth. Aircraft pilots operating under instrument flying conditions in clouds or at night can fall victim to vertigo.

Vertigo can be and is deliberately created on the ground by a rotating seat known as a Barany chair. It's used to acquaint aviators with vertigo because the danger of vertigo can be reduced if you understand the nature and causes of the condition, avoid the environmental situations that cause it and learn to heed and *believe* visual inputs from vehicle instruments that monitor and report on the aircraft's attitude and motion.

Once you've experienced vertigo, there's no question about what it's like. And anyone who's experienced it agrees that once is enough! If you know what it is, you can prevent it from happening to you again.

Long-Term Weightlessness Effects

Although more than two decades of human space experience have shown that you can probably adapt to weightlessness and readapt to living in Earth's gravity again upon your return, all of the answers aren't known yet for long-term space living in weightlessness. We don't yet know what will happen to you if you elect to spend your life in the weightlessness of space. Some of the early data from flights up to six months' duration are disturbing.

Soviet cosmonauts have spent far longer periods of time in weightlessness than have American astronauts. But the data from Soviet missions isn't as widely available as those from American sources and we don't really know the precise physiological data from cosmonauts who've spent up to six months in space and have returned to space for a second lengthy stay. The longest period spent in weightlessness to date by Americans was slightly more than eighty-four days in the case of the *Skylab 4* crew: Gerald P. Carr, William R. Pogue, and Edward G. Gibson.

The data from the *Skylab 4* mission indicates that there were significant changes in body chemistry and function experienced by this crew.

From this data, we believe now that most of the physiological changes to your body caused by weightlessness *appear* to be reversible—i.e., once you return to Earth, the changes in your body caused by weightlessness disappear and you become just like everyone else again.

But the physiologists who've studied the Skylab data aren't certain. And they don't know whether or not very long stays in weightlessness would indeed lead to irreversible changes in your body.

Any part of your body that isn't used on a regular basis will eventually decay or atrophy. As might be expected, your bones and muscles don't have as much work to do supporting your body in weightlessness compared to what's required of them in Earth's gravity field. After you've been in weightlessness for months, you might find that your muscles had lost strength and tone, causing difficulties in moving around after a return to Earth's gravity. The degree to which you might prevent this by a program of extensive exercise in space isn't known at this time.

Therefore, your bones and muscles will begin to show changes after a short stay in weightlessness. But bone decalcification appears to be a major problem with extensive living in weightlessness. Your bones aren't required to be as strong in weightlessness and therefore begin to lose the calcium that contributes to their strength.

Bone decalcification in weightlessness occurs at a rate of between 1 and 2 percent of your bone mass per month, resulting in decreased bone mass and density. The eighty-four-day *Skylab 4* mission data indicated that this bone-decalcification problem doesn't diminish with time. And the consequences or effects of this are *really* troublesome.

The Effects of Calcium Resorption

When your bones lose calcium, it's released into your body fluids, primarily the blood serum—because this is where your body gets its supply of calcium under normal conditions. Ninety-nine percent of your body's calcium is in your bones, and 1 percent of this bone calcium is freely exchangeable with extracellular body fluids. Your normal blood-calcium level is 8.8 to 10.4 milligrams per 100 millimeters of blood. Forty percent of this is chemically bound to serum proteins in your blood.

An excess of calcium in your blood is known as *hypercalcemia*. It occurs when your blood-calcium level exceeds 10.5 milligrams per 100 milliliters. On Earth, hypercalcemia can be caused on a temporary basis by excessive ingestion of foods having high calcium content, by cancer, or by endocrine imbalances.

We don't yet know if you'll suffer from severe hypercalcemia because of spending long periods of time in weightlessness.

Symptoms of mild hypercalcemia include constipation, nausea, vomiting, abdominal pain, and often the urge for nearly constant urination. None of these mild hypercalcemic symptoms have been reported by any astronauts. We don't know whether or not Soviet cosmonauts have reported such symptoms during or after their longer stays in space.

When your blood-calcium level exceeds 12 milligrams per 100 milliliters, severe symptoms such as confusion, delirium, psychosis, stupor, and coma are possible. When it exceeds 18 milligrams per 100 milliliters, the results include shock, liver failure, and death. None of the severe hypercalcemic symptoms have been reported by astronauts, and the data from Soviet cosmonauts isn't in hand. The mild forms of hypercalcemia that have been measured in astronauts thus far have shown up as some imbalance of the electrolytes in the blood as well as imbalances in some important hormone outputs and proportions. There are some indications that these hormonal imbalances created unstable protein

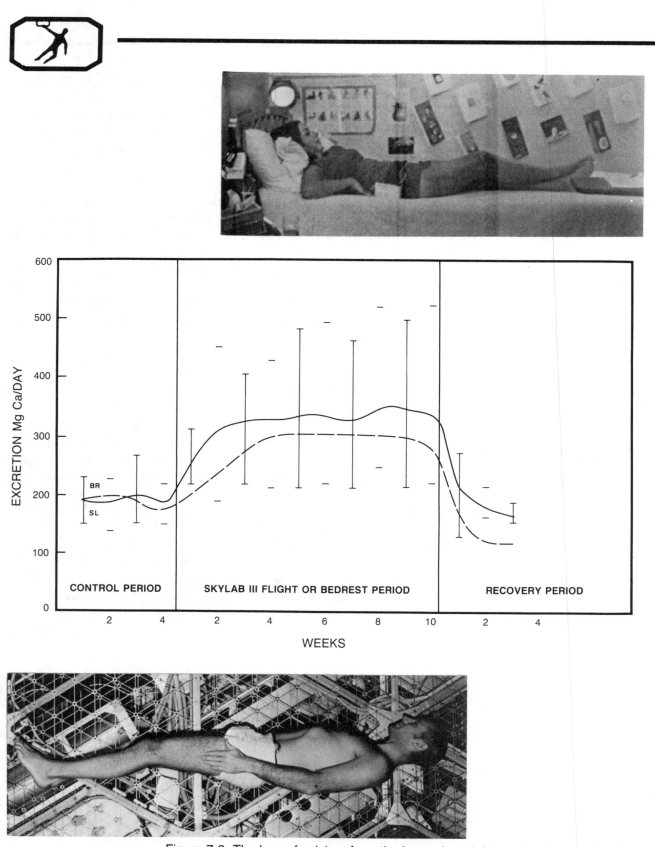

Figure 7-3. The loss of calcium from the bones in weightlessness is very closely analogous to that lost on Earth during prolonged bed rest. *(NASA)*

and carbohydrate states, hypoglycemia, and unusual increases in both primary and secondary pituitary hormone levels. These are the sorts of things doctors expect in cases of hypercalcemia here on Earth.

It's interesting to note that a symptom of this hypercalcemic condition is the development of highly active neural processes due to an increase in a hormone known as norepinephrine. This substance acts to encourage the chemical transmission of nerve impulses across the synaptic junctions between nerve cells. There has been little research on any possible connection between this hormonal change and the changes in thought patterns experienced by many astronauts, including "existential" or religious experiences some of them have reported both in orbit and on the surface of the Moon following a four-day flight in weightlessness.

On Earth, hypercalcemia is a metabolic or mineral imbalance disorder that can be treated with immediate results in most cases by means of steroid therapy using a drug called prednisone. Long-term treatment of hyper-calcemia involves the oral administration of phosphorus. Whether or not the terrestrial therapy for hypercalcemia can be used to treat you if you begin to exhibit the symptoms and etiology of hypercalcemia remains to be seen. All of the data isn't in, but the calcium-resorption problem has medical experts concerned. They'll know much more about this in a few years. Although it is a matter of concern for those who would like to spend long periods of time in weightlessness, the fact that there's an existing therapy bodes well for the short-term solution to the problem. It is highly probable that the calcium-resorption problem will have a solution by the time our physical technology has reached the point where it is possible to build very large space colonies and support lunar and planetary facilities. But someone is going to have to start searching for the solution soon if they aren't working on it already.

Cardiovascular Changes

There have been a few other physiological problems encountered by astronauts in long-term space missions to date. The major ones are also based on the fact that the human species evolved in Earth's gravity. The absence of that gravity force has caused changes in the heart and cardiovascular system of all astronauts and cosmonauts.

In Earth's gravity, your heart and its system of blood vessels must contend with the problem of pumping blood up from your legs and feet while at the same time maintaining the proper blood pressure in your brain, which is located above your heart. Weightlessness upsets this dynamic system, but your body appears to adapt quickly to it.

On return to Earth, however, postflight medical examinations of astronauts have revealed an increase in heart rate of 10 to 20 beats per minute, changes in muscle reflexes, and pooling of the blood in the lower abdomen and legs.

Medical researchers are concerned that long-term living in the zero-g of space may lead to irreversible changes in your cardiovascular system coupled

Human beings seem to have had very little trouble adapting to zero-g—and enjoying it, too. (NASA)

with a decreased effectiveness in your body's resistance to disease because of changes in your blood chemistry itself.

All of these physiological reactions to weightlessness and their consequences may mean that your stays in weightlessness might be limited to six months or less; if you decide to live permanently in the weightless condition in space, you may not be able to return to Earth because of both the gravity field and your increased susceptibility to disease. You may see space medicine develop therapies for these conditions. Or we'll have to design and build space habitats with slowly spinning wheel-like living portions where centrifugal force can act as pseudogravity to prevent these weightless syndromes, symptoms, and problems from occurring. You'd then be able to spend short periods of time

110

actually working in weightlessness and then return to the centrifuged living module for meals, recreation, and sleep.

However, please keep in mind that the data isn't complete yet. Although Soviet literature on the subject of human tolerance and adaptation to weightlessness is not widely available in the West, NASA data on stay times up to eighty-four days seems to indicate that you can indeed function effectively for long periods of time in weightlessness and then return to Earth without experiencing adverse effects. The physiological changes in your body caused by adaptation to weightlessness and created by the weightless condition have thus far proved to be reversible.

And weightlessness seems to be one of the things aspiring space colonists look forward to experiencing. Astronauts have found it's a very pleasant experience. Said astronaut Dr. Joe Kerwin of the *Skylab 2* crew, "It was a continuous and pleasant surprise to me to find out how easy it was to live in zero-g, and how good we felt."

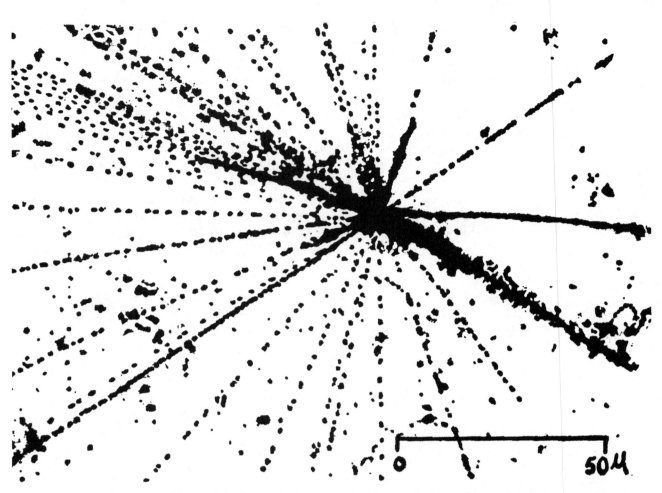

Figure 8-1. When high-energy ionizing radiation collides with atoms or molecules, fission or destruction of the molecule can occur.

8 Radiation

Space was once considered to be a quiet place, a vacuum in which there were local concentrations of matter known as stars, planets, planetoids, moons, comets, and meteors. Then a new factor entered the picture: ionizing radiation of the sort that on Earth comes from X rays, radium, and other radioactive sources. Scientists detected "cosmic rays," which were a form of very energetic ionizing radiation that seemed to come from all directions in space.

But no one really anticipated what they'd find in the way of space radiation when the first United States earth satellite, *Explorer-I*, went into orbit on January 30, 1958. *Explorer-I*'s instruments, designed by Dr. James Van Allen of the State University of Iowa, were to detect and measure the amount of radiation in orbit. The experiment was carefully designed using instruments similar to Geiger counters whose measurement ranges were determined on the basis of other cosmic-ray measurements made over a few places on Earth—White Sands, New Mexico, and Fort Churchill, Canada, to name but two—at altitudes up to 150,000 feet using balloons, and at higher altitudes but over shorter periods of time by rockets. *Explorer-I* provided the first chance to gather cosmic-ray data from an experiment that would stay in space for months.

The Van Allen Belts

Explorer-I ran into a storm of radiation that overloaded Dr. Van Allen's instruments. There had been no reason to believe there was such intense radiation in

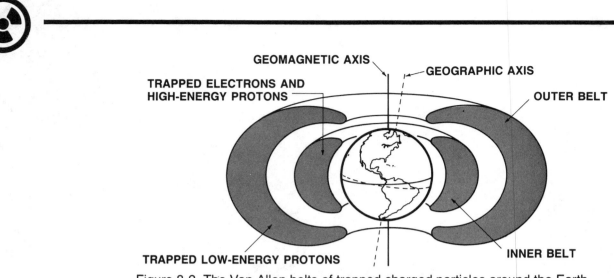

Figure 8-2. The Van Allen belts of trapped charged particles around the Earth *(Art by Sternbach)*

space. Van Allen redesigned his equipment using radiation counters with high saturation limits. It was launched in *Explorer-III* on March 26, 1958, and reported the exact radiation level. Subsequent satellite flights and space probes mapped these radiation belts that surround the Earth and bear Dr. Van Allen's name today.

An understanding of the nature of the Van Allen radiation belts is extremely important if you are to stay alive and healthy in space. Not only do the Van Allen belts offer you some protection against space radiation if a habitat is orbiting between the belt and the Earth, but the presence of the belts themselves opens the extremely important subject of ionizing radiation in space. Before Dr. Van Allen's discoveries, no one had any inkling of the unseen hazard you'd have to contend with in space. Now space radiation is one of the most important elements in space habitation because of the effects of such ionizing radiation on your health.

The Van Allen belts exist because the Earth has a magnetic field and orbits

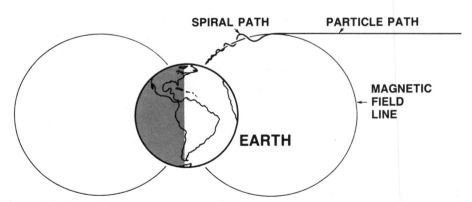

Figure 8-3. Incoming nuclear particles from the solar wind are captured by the Earth's magnetic field and spiral around the magnetic lines of force back and forth between the Earth's magnetic poles. *(Art by Sternbach)*

114

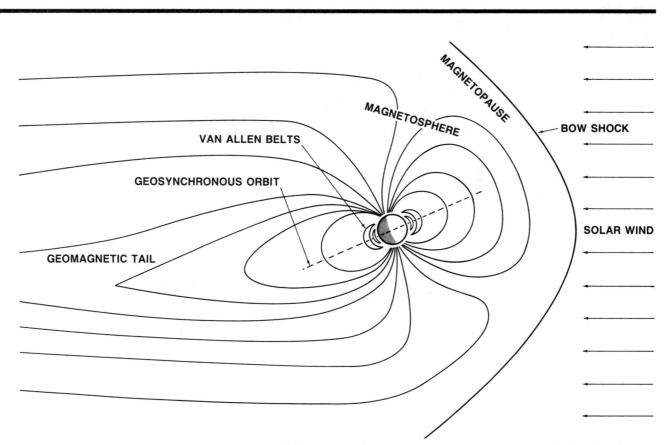

Figure 8-4. The Earth's magnetosphere interacts with the solar wind, deflecting most of the solar wind particles. *(Art by Sternbach)*

in the atmosphere of a star, the sun. The Earth is one of the few planets that possesses a strong magnetic field, which exists, according to the latest theories, because the Earth probably has a molten nickel-iron core. Eddies caused by rotation of this molten core produce small magnetic fields. When the fields of all eddies are combined, they add up to a large terrestrial magnetic field with lines of force that swoop out into space from the magnetic poles that are near but not identical to the poles of rotation. This large magnetic field is not neatly symmetrical like the field of a small bar magnet, but instead is distorted by the magnetic field of the sun and the charged particles blasted into space by the sun.

The Solar Wind

The hydrogen-fusion energy process of the sun produces charged particles—negatively charged electrons and positively charged protons. They stream out from the sun as part of the "solar wind." During the Earth's journey through space around the sun at an orbital speed of some 18 miles per second, it runs into enormous quantities of these charged atomic particles ejected from the sun.

Since electrons and protons have an electric charge, they're deflected from

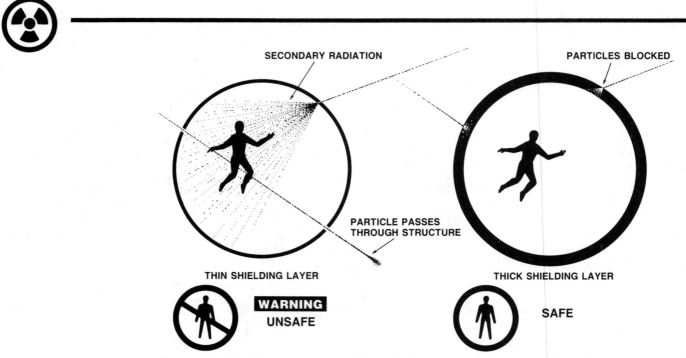

SECONDARY RADIATION

PARTICLES BLOCKED

PARTICLE PASSES THROUGH STRUCTURE

THIN SHIELDING LAYER

THICK SHIELDING LAYER

WARNING UNSAFE

SAFE

Figure 8-5. When an electron collides with the metal hull of a spacecraft, it releases X rays by the *bremsstrahlung* phenomenon. *(Art by Sternbach)*

their original courses when they encounter the Earth's magnetic field. Some of them become trapped in the terrestrial magnetic field and circle back and forth between the Earth's magnetic poles along the magnetic lines of force. Some of them manage to escape again into space. Others encounter the Earth's upper atmosphere near the poles where, if the sun is active and putting out a lot of particles, these trapped solar particles interact with the upper atmosphere to create the auroras.

In space, this cloud of trapped solar particles forms doughnut-shaped belts around the Earth roughly over the equator. The density of the charged particles in the belts reaches several maximums, from where the belts begin about 100 miles up to where they finally disappear about 17,000 miles above the Earth's surface.

An electrically charged particle can create X rays or gamma rays by a process called *bremsstrahlung*, a German scientific word meaning "radiation from slowing down." To understand bremsstrahlung, which is an important part of space living, you must make an imaginary visit to the local hospital.

X Rays

In 1895, William Roentgen of Germany was experimenting with a glass bottle with electrodes at both ends and a vacuum inside. The device is called a Crookes tube because it was developed by Sir William Crookes in 1879. When Roentgen applied several thousand volts of electrical potential between the electrodes of a Crookes tube, he generated a new form of electromagnetic radiation that would pass through human bodies and, if a photographic plate were exposed to the result, would reveal the bones and other innards of the body thus irradiated.

Today's hospital X-ray machine still uses a Crookes tube. One electrode of the tube contains a heated filament or cathode from which electrons are "boiled off" in the same manner as inside the TV picture tube (yet another form of the Crookes tube). At the other end of the tube is a positive electrode, the anode, which is charged several thousand volts positive with respect to the heated cathode. The electrons leave the hot cathode and accelerate rapidly in the high-voltage electric field between the cathode and the anode, acquiring enormous energies as they travel. These energies are expressed in terms of "electron volts."

An electron volt is defined as the amount of energy given to a particle with a unit electric charge when it's accelerated through a potential difference of one volt. Because of the very small value of this unit, it is most usually expressed in multiples such as "million electron volts," or MeV.

The greater the voltage differential between the electrodes of a Crookes tube or a TV picture tube, the greater the electron speed and therefore the higher the MeV energy. The electrons finally slam into the anode, and their energy is converted into X rays, which emanate from the anode.

Any electronic vacuum tube with several thousand volts across it will generate X rays. In fact, a TV picture tube generates low-energy or "soft" X rays, but the level of ionizing radiation thus produced is very low.

Therefore, ionizing radiation can be generated in any space facility traveling through space at several miles per second because it collides with protons

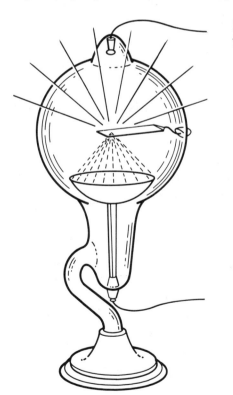

Figure 8-6. The original form of the Crookes tube *(Art by Sternbach)*

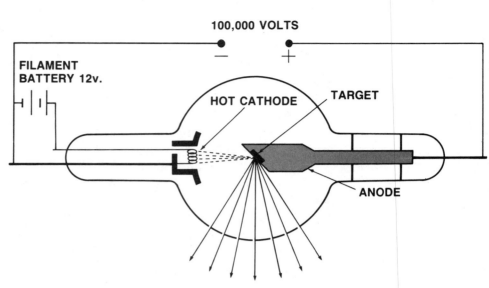

Figure 8-7. The modern X-ray tube is an improved version of the Crookes tube and works the same way. *(Art by Sternbach)*

or electrons of the solar wind. When these charged particles slam into the metallic outer hull of a space facility or spacecraft, this duplicates the action of the Crookes tube, and radiation is produced by bremsstrahlung.

Bremsstrahlung radiation is only one form of ionizing radiation that is present in space. There are other forms and sources of radiation, some of which are exceedingly deadly.

Other Ionizing Radiation

Ionizing radiation is of two sorts: high-energy photons from the electromagnetic spectrum; and high-velocity, and therefore high-energy, charged particles.

The electromagnetic (e-m) spectrum is shown in Figure 8–8. E-m radiation is characterized by its wavelength or frequency—the two are directly related, the higher frequencies having the shorter wavelengths. Because e-m radiation can behave as though it were composed of particles of negligible mass, e-m radiation is considered to consist of "photon" particles. The higher the frequency, the greater the photon energy expressed in terms of MeV.

The low-energy or low-frequency end of the e-m spectrum contains the radio waves starting with the long-wavelength waves at 11.8 to 13.1 thousand cycles, or kilohertz, frequency—the frequency primarily used today for the Omega radio navigation system on Earth. The AM and FM radio broadcast bands are of higher frequency—0.5 to 1.6 million cycles, or megahertz (mHz) for AM and 88 to 108 mHz for FM. Above the radio-frequency spectrum are the microwaves, which include those frequencies used for radar. These blend into the infrared portion of the spectrum, which, in turn, blends into the visible spectrum that we detect as colors, red being the longest light wave-

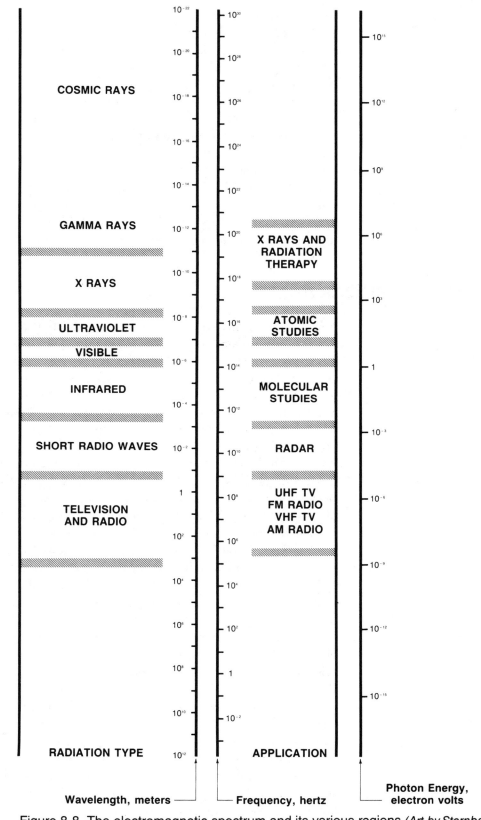

COSMIC RAYS

GAMMA RAYS

X RAYS

ULTRAVIOLET

VISIBLE

INFRARED

SHORT RADIO WAVES

TELEVISION
AND RADIO

RADIATION TYPE

X RAYS AND
RADIATION
THERAPY

ATOMIC
STUDIES

MOLECULAR
STUDIES

RADAR

UHF TV
FM RADIO
VHF TV
AM RADIO

APPLICATION

10^{-22} 10^{30} 10^{15}
10^{-20} 10^{28}
10^{-18} 10^{26} 10^{12}
10^{-16} 10^{24}
10^{-14} 10^{22} 10^{9}
10^{-12} 10^{20} 10^{6}
10^{-10} 10^{18} 10^{3}
10^{-8} 10^{16}
10^{-6} 10^{14} 1
10^{-4} 10^{12}
10^{-2} 10^{10} 10^{-3}
1 10^{8} 10^{-6}
10^{2} 10^{6}
10^{4} 10^{4} 10^{-9}
10^{6} 10^{2} 10^{-12}
10^{8} 1
10^{10} 10^{-2} 10^{-15}
10^{12}

Wavelength, meters   Frequency, hertz   Photon Energy, electron volts

Figure 8-8. The electromagnetic spectrum and its various regions *(Art by Sternbach)*

119

length and violet being the shortest. Immediately above this is the ultraviolet portion of the spectrum. Then the X-ray portion, capped by the gamma-ray segment of the spectrum. On the upper end of the e-m spectrum are the cosmic rays, which have the shortest wavelengths, highest frequencies, and highest-energy photons.

Particle radiation is primarily composed of either electrically charged or neutrally charged subatomic particles—the negatively charged electron, the positively charged proton, and the neutral neutron—or high-velocity ionized atoms that have had their outer shell electrons stripped off and therefore carry an electric charge. Although the number of different kinds of nuclear and subnuclear particles keeps growing constantly as physicists probe the basic nature of matter and seek to understand it, the electron, proton, and neutron suffice to characterize the radiation of space. The energies of these particles are also expressed in terms of MeV.

Your Radiation Sensors

You're wonderfully equipped to detect things that go on about you. Evolutionary processes over the past several billion years have equipped life forms on Earth with sensing devices to detect danger. You're well provided with natural sensors to detect pressure, sound, light, heat, acceleration, form, color, and so forth. These are shown in Figure 8–9 along with the sensations they produce and the interpretation you give these sensations. Your sensors for the electromagnetic spectrum are excellent for covering the survival factors with which your hunting ancestors had to contend, but their sensitivity and range are limited. In the face of modern nuclear technology and our thrust into the new environment of space, these limitations can get you into serious trouble because you can't sense some regions of the e-m spectrum that have definite effects upon your physical health and well-being.

You'll have excellent sensors for light radiation: your eyes. You can detect infrared or heat radiation with your skin. You can detect the *results* of overexposure to ultraviolet radiation because it causes sunburn. But you can't detect radiation in other parts of the e-m spectrum. (If you had natural sensors for the radio part of the e-m spectrum, you could tune in a radio station without using the electronic crutch of a transistor radio.)

It wasn't known until World War II that human beings could see ultraviolet, or u-v, radiation, but research in Great Britain indicated that a few airplane pilots could see u-v. It was therefore proposed that some of the airfields of the Royal Air Force be illuminated at night with u-v so that these unusual pilots could see to land their bombers. However, it turned out that the type of people who could see u-v best were blond, blue-eyed, Teutonic types—and there were plenty of pilots like that in the Luftwaffe.

Electromagnetic radiation of all types has some effect upon you. The physiological effects of some parts of the e-m spectrum are well known—sunburn, for example. But the effects of other types of radiation aren't yet known or even understood. And a great deal more is misunderstood and therefore feared.

Human Stimulus-Receptor-Sensation Relationship

Stimulus	Receptor	Sensation	Interpretation
E-m waves, 10^{-5} to 10^{-4} cm	Photoreceptor	Light, color	All visual impressions
E-m waves, 10^{-4} to 10^{-2} cm	Skin thermo-receptors	Temperature hot to cold	Heat, fire, warmth, comfort, freezing
Mechanical oscillations 20-20,000 Hz	Inner ear, cochlea	Noises, sound	Sounds, voices, music tones
Pressure	Tangoreceptors of the skin and pressoreceptors in the body	Touch, weight, acceleration, deceleration	Objects, liquids, states of motion
Linear accelerations	Otoliths	Changes, state of motion, equilibrium	Movements, voluntary and involuntary motion
Angular accelerations	Semicircular canals	Rotation	Turning, voluntary and involuntary rotations
Chemicals in watery solution	Taste buds	Sweet, bitter, sour, salt	Food
Chemicals in gaseous state	Olfactory cells	Smells	Odors of certain substances
Chemical and mechanical inner changes	Proprioceptors of the muscles and connecting tissue	Internal tension and pressure	Hunger, thirst, digestive processes
High-energy effects of all kinds	Free nerve endings of sense organs	Pain	Extreme stress, injury, illness

Figure 8-9.

This is particularly true of the neurological and psychological effects of radio waves and microwaves, for example, although new information is becoming available all the time. Twenty-five years ago, nobody thought there were any physiological or psychological effects caused by exposure to radio- or microwave radiation, but the continuing influx of data now seems to indicate beyond a shadow of a doubt that *something* is happening. Radio waves up to a frequency of about 500 mHz (wavelength about 30 centimeters) don't appear to have serious physiological effects upon organic life except at very high energy levels, where there *may* be some neurological and therefore psychological effects. Some research is now going on in this area in both the United States and the Soviet Union.

Microwaves are radio waves with frequencies of 1,000 mHz or more. Microwaves at certain frequencies and power levels can be harmful to living things because organisms contain a large amount of water. The water molecule

121

is resonant to microwave energy at about 2,450 megahertz and will therefore absorb microwave energy at that frequency and convert this energy to molecular vibration, which is manifested as heat. A microwave oven works because its e-m radiation vibrates the water molecules in foods and therefore heats food. Microwave ovens are carefully shielded to prevent the escape of the microwave radiation inside. But you can't sense the presence of microwave energy, only the effect of that energy: heat. By the time you detect it, it may be too late to do anything about it. The long-term effect of exposure to high-energy microwave radiation at very low power levels is still a subject of intense study.

You can detect infrared radiation as heat from a potbellied stove, a radiant heater, or the sun.

You can also detect light radiation.

You can detect ultraviolet radiation only after you've gotten the sunburn. The short-term effects of u-v radiation on your body cause sunburn and eventually a deep tan as your skin acts to protect itself by concentrating more of the pigment melanism in its upper layers to reflect the u-v. Some types of people whose ancestors evolved for many generations in the sunny parts of Earth are naturally protected by a hereditary concentration of melanism in their skin. The long-term effects of excessive ultraviolet radiation appear to cause skin cancer in light-skinned people by a mechanism not yet well understood by researchers. However, as one cancer researcher pointed out, overstress *any* organism by *any* means, and it will probably develop some form of cancer.

The Physiological Effects of Radiation

At frequencies and wavelengths in the ultraviolet and above, the energies of the photons are high enough to create damage in living organisms. The mechanism by which this is done is well known to scientists and technologists, but not well understood by the layman. Therefore, since people can't sense high-energy radiation in the ultraviolet, X-ray, and gamma-ray portion of the e-m spectrum, laymen have a tendency to harbor an unreasoning fear of it. The word "radiation" itself has a highly negative semantic content—it invokes fear and apprehension. In this book, the word "radiation" has been used in its precise scientific context thus far. There's nothing to fear from radiation per se; some types of radiation are actually beneficial to human beings. Who would want to go around in a world without light radiation, for example? Or infrared radiation?

The *energy* content of radiation is the factor that can harm living organisms. The higher the frequency of the photon, the higher its energy.

When a photon with radiowave MeV energy levels goes through a living cell (as uncountable numbers of them do every day), it creates practically no damage whatsoever because it doesn't have enough energy to do damage. At microwave frequencies, the photon causes damage because it excites the water molecules present in all living cells. At infrared energies, it excites all atoms and molecules, creating heat. At light wavelengths, its energy can be chemically detected only by special, highly evolved cells in the retina of the eye, but a

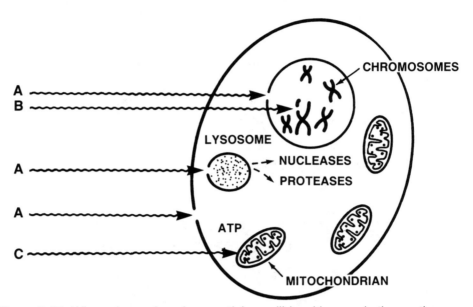

Figure 8-10. When charged nuclear particles collide with organic tissue, the energy strips away electrons from the tissue molecules, ionizing them and changing their chemical characteristics. *(Art by Sternbach)*

light-energy photon doesn't do any cellular damage unless it's the sort of intense, coherent light that comes from a laser.

However, when the energy level of a photon puts it in the ultraviolet and X-ray portion of the spectrum, a photon can indeed cause microscopic damage inside a living cell.

When a high-energy photon plows into a living cell, the matter in the cell slows it down and makes it give up some of its energy. It does this by ionizing some of the atoms in the cell, stripping them of their orbital electrons, and leaving them with an electric charge. These ionized atoms don't behave the same in a chemical sense as their nonionized companions. Their presence disrupts the delicate atomic and molecular balances within the cell, changing it so that it doesn't work the same as before.

Measuring Radiation

There's been a great deal written about ionizing radiation—much of it "scare" writing that emphasized the hazards. Perhaps some rationality can be introduced into the subject by recalling the words of Lord Kelvin on the importance of numerical measurement in scientific thinking quoted earlier.

The various types of radiation can be measured, and the relative biological damage that can be caused is also known.

The most common measure of radiation dosage is the *rad*, which stands for "radiation absorbed dose." This is the amount of energy dissipated by radiation in matter. One rad equals 0.0000024 calories per gram. This isn't very much, but we're not talking about big things.

123

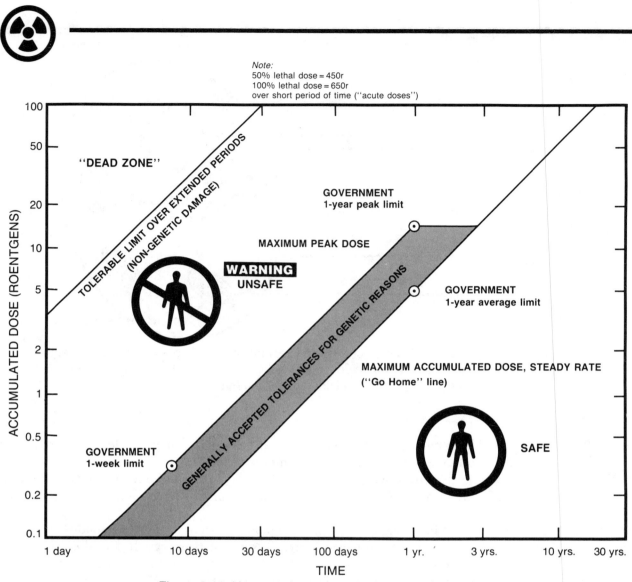

Note:
50% lethal dose = 450r
100% lethal dose = 650r
over short period of time ("acute doses")

Figure 8-11. Human tolerances to long-term exposure to radiation *(Art by Sternbach)*

A similar unit called the *roentgen*, named after the nineteenth-century scientist who discovered X rays, is also a measure of exposure dose. One roentgen of gamma radiation produces 87 ergs of energy per gram of mass. An erg is the amount of energy required to move one gram of mass one centimeter—again, not a large value.

There's also a unit called the *rem*, standing for "rad equivalent, mammal." It's a unit reflecting the ability of radiation to produce damage in the tissue of a mammal.

The *dose rate* is the radiation received per unit of time. Detectable biological effects increase as the dose rate increases. Roughly stated, an observable effect is certain with dose rates greater than 4 rads per minute.

Body area exposed is also a factor. The entire human body can absorb up to 200 rads without fatality. However, as the whole-body dose approaches 450 rads, the death rate becomes about 50 percent. A whole-body dose of more than 600 rads in a short time is always fatal. But many thousands of rads can be delivered to a human body for cancer therapy over a long period of time if only small portions of the body are irradiated. And if the bowels or the bone marrow are adequately protected by shielding, an individual may survive what might otherwise be a fatal whole-body dose.

Natural-Radiation Dose

There's a gentle, low-intensity rain of high-energy radiation around you all the time on Earth. It comes from both the Earth itself and from space. The Earth-source radiation comes from naturally radioactive elements that are present in nearly all rocks and soils. The space-source radiation comes from the sun and from the galaxy. Organisms on Earth have adapted to living under this continual bombardment of natural radiation; if it were taken away by any attempt on the part of people to irrationally create a "pure" environment without any "bad radiation," nobody really knows what the long-term effects on the Earth's living organisms would be.

The cells of every human body are also subjected to this natural or "background" radiation. They die and replace themselves regularly. Because of this, you're not the same person you were last year. Human beings have developed an immunity to this background radiation.

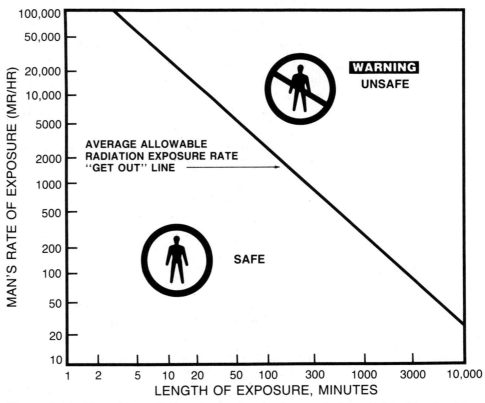

Figure 8-12. Allowable radiation dose rates for human beings *(Art by Sternbach)*

NASA Operational Flight-Crew Exposure Limits

(All exposure units expressed in rems)

Constraint	Bone Marrow	Skin	Eye
Thirty-day maximum	25	75	37
Quarterly maximum	35	105	52
Yearly maximum	75	225	112
Career limit	400	1,200	600

Figure 8-13.

Each year, the average human being is subjected to the following dose rates from natural sources in terms of millirads per year (mr/yr):

Natural radioactive material in the bones: 34 mr/yr

Cosmic rays (at sea level): 30 mr/yr

Natural radioactivity of the surroundings: 48 mr/yr

Medical X rays: 75 mr/yr

Radioactivity from man-made sources: about 12 mr/yr

Total: about 200 mr/yr average

You can handle this natural background radioactivity without any discernible physiological effects. In fact, the development of life on Earth may have depended upon this background radiation to create genetic mutations, which in turn caused the variation as well as the development of species such as *Homo sapiens*.

According to the National Academy of Sciences, you can absorb up to 10 rads in a thirty-year period without danger—you normally get about 6 rads from background radiation in that time period. Human physiology can probably handle even more than that. Nuclear workers are permitted to receive dosages of 300 millirads per week for a total of 46.8 rads in a thirty-year period.

Figure 8–13 shows the operational exposure limits in rems permissible for Space Shuttle flight crews in space.

Heavy Radiation Exposure

When you get too much high-energy ionizing radiation, so many of your cells may be affected that they can't replace themselves. Or it may affect cells in such a way that their natural workings are disturbed, converting them into rampaging, ever-growing, all-consuming cancer cells instead.

When subjected to intense levels of ionizing radiation, far too many cells are damaged, and radiation sickness is the result.

The physiological effects of radiation are now well known. Your body tissues vary in response to immediate irradiation in the following order of sensitivity, the most sensitive being listed first:

1. lymphoid cells
2. the reproductive cells of the gonads
3. bone-marrow cells
4. epithelial cells of the intestines
5. the epidermis
6. hepatic (liver) cells
7. the epithelium of the lungs
8. kidney tissue
9. cells of the intestinal cavity
10. nerve cells
11. bone cells
12. muscles and connective tissues

Large but sublethal doses affect both the rate of mitosis (cell division) and the synthesis of DNA. This can lead to diminished production of new cells in tissues that normally undergo continual renewal—bone marrow and gonads, for example. Some cells are so badly damaged that they may continue reproduction but produce abnormal progeny. Some of these altered cells may be cancerous.

A rule of thumb, then, for exposure to ionizing radiation is:

A little is probably necessary, but a lot over a short period of time isn't. A lot of it over a long period of time may or may not be harmful, depending where it was absorbed by your body.

Radiation in Space

Although you can handle the natural radiation here on Earth—and to a large extent even the unnatural radiation from nuclear power plants and other man-made sources if dose rates are low enough—measurements made of the type and amount of radiation in space indicate that this unseen hazard definitely will pose a problem for you in space.

Figure 8–14 shows the types of ionizing radiations that exist in space above the Van Allen belts, their energy levels in MeV, the "relative biological effectiveness" (known as RBE) of each, and where each type comes from.

Because different forms of radiation with different energies may have different effects upon organic tissue, the radiation dose from each must be multiplied by the RBE. The product is then a measure of the danger of the

Ionizing Radiations in Space

Name	RBE	Source
X rays and gamma rays	1	Radiation belts, solar radiation, and bremsstrahlung electrons
Electrons		
1.0 MeV	1	Radiation belts
0.1 MeV	1.08	
Protons		
100 MeV	1–2	Cosmic rays, inner-radiation belts, and solar cosmic rays
1.5 MeV	8.5	
0.1 MeV	10	
Neutrons		
0.05 eV (thermal)	2.8	Nuclear interactions in the sun
0.0001 MeV	2.2	
0.005 MeV	2.4	
0.02 MeV	5	
0.5 MeV	10.2	
1.0 MeV	10.5	
10.0 MeV	6.4	
Alpha particles		
5.0 MeV	15	Cosmic rays
1.0 MeV	20	
Heavy primaries	*	Cosmic rays

*Damaging power of heavy primaries varies widely and is measured in terms of how many chemical bonds per unit of body mass are broken, thereby giving an indication of the tissue damage sustained.

(From "Space Settlements, A Design Study," National Aeronautics and Space Administration publication SP-413, Government Printing Office, 1977.)

Figure 8-14.

particular kind of radiation and is usually presented in terms of rems. Thus, a low-energy thermal neutron would have 2.8 times the effect of a 1-rad exposure of X rays.

X rays and gamma rays come from the Van Allen radiation belts by bremsstrahlung electrons and protons as well as from nuclear reactions.

Electrons come from the sun in the solar wind and are the most numerous charged particles trapped in the Van Allen belts. Their primary hazard is their ability to produce bremsstrahlung radiation in the X-ray and gamma-ray region of the e-m spectrum.

Protons also come from the sun in the solar wind and are also trapped in the Van Allen belts. They're also part of the overall high-energy radiation called cosmic rays.

Neutrons are produced by nuclear interactions. The primary source for these in the solar system is the thermonuclear fusion process of the sun.

Alpha particles are part of cosmic radiation and, even at low energies, have large RBE values.

The "heavy primaries" listed in the table are the major constituents of cosmic radiation.

Cosmic Radiation

Cosmic radiation has been a source of much frustration to nuclear physicists and appears to be composed mostly of high-energy atomic nuclei or atoms that have had all their orbital electrons stripped away. About 80 percent are hydrogen nuclei or protons; 16 percent are helium nuclei consisting of two protons and two neutrons with an electric charge of +2; and 4 percent are made up of the nuclei of heavy atoms such as iron. Cosmic-ray energies are extremely high, ranging from 100 MeV to several thousand MeV. The source of cosmic rays is unknown, although some of them probably originate in the sun while the remainder may originate elsewhere in the galaxy, because they appear to come from all directions in space.

The Earth's atmosphere acts as an excellent shield: only a small percentage of the overall cosmic radiation reaches the ground. The atmosphere's effect on cosmic radiation is somewhat like the effect of a dense forest on a rifle bullet: The bullet won't go very far before it hits a tree. The heavier cosmic-ray primaries collide with atmospheric molecules at high altitudes, while the lighter and smaller particles penetrate farther into the atmosphere before they suffer a collision. When they do, nuclear fireworks go off.

When a cosmic-ray particle, especially one of the slow-moving but heavy primaries, hits an atmospheric molecule, the release of this high energy causes

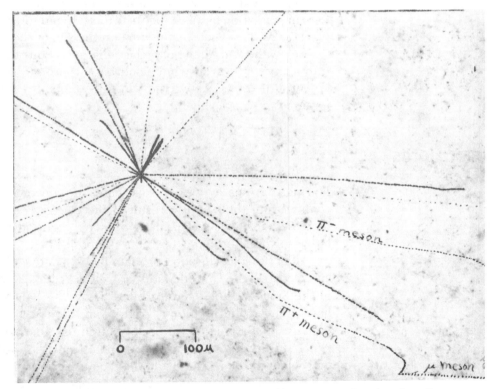

Figure 8-15. When nuclear particles collide with atoms and molecules, they create a spray of "secondary" particles or radiation such as shown in this historic photo of a cosmic-ray collision, made by Dr. Herman Yagoda in 1955. (Dr. Herman Yagoda)

a nuclear disintegration with protons, electrons, neutrons, gamma rays, and other nuclear debris scattering in all directions. The spray of nuclear junk resulting from the collision of a heavy primary with an atmospheric molecule is known as "secondary cosmic radiation." The ionizing energy of these secondaries is appalling within several millimeters of the occurrence.

Secondary cosmic radiation is around us all the time on the surface of the Earth, but it isn't harmful because a cosmic-ray particle must have an energy of more than 1,000 MeV to reach the ground at the Earth's equator. Secondary cosmic radiation can be observed at an increasing intensity as you ascend into the atmosphere. However, at altitudes above about 75,000 feet, nothing but primary cosmic-ray particles exist because there isn't enough atmosphere to cause collisions and secondaries.

The light, fast, and most common cosmic-ray particles are the protons (hydrogen nuclei or ionized hydrogen) and alpha particles (helium nuclei or ionized helium). They have from 1 to 20 times the ionizing potential of X rays, depending upon their energies, but they have a tendency to pass through organic tissue so rapidly that there's limited time to cause ionization. The slow, heavy primaries are the worst, however, because they plow into organic matter and expend their energies by leaving a broad trail of ionization that kills cells. When it comes to cosmic rays, the ionizing power increases as the particle energy decreases.

Figure 8–14 doesn't indicate the relative population or concentration of ionizing radiation in space, because this depends on where the measurement is taken as well as what's happening on the sun at the moment.

Below the Van Allen belts, the radiation is quite different, because the belts trap many of the solar-wind electrons and protons, thus acting as a shield for low-orbit habitats. Above the Van Allen belts, the full force of the solar wind impinges on a space habitat.

Solar Flares

During an event known as a solar flare, the concentration of charged particles coming from the sun as a result of the energy release of the solar flare can increase enormously. This can produce a high-radiation environment that is lethal to you after only a few hours' exposure.

All of the long-duration orbital flights to date—Mercury, Gemini, Apollo, and Skylab in the United States, as well as Vostok, Voskhod, Soyuz, and Salyut in the Soviet Union—have orbited below the Van Allen belts and therefore have enjoyed the protection of the Van Allen belts, which trap the charged particles that stream out from the sun in the solar wind.

To ensure the safety of the Apollo astronauts, the flights to the Moon not only went through the Van Allen belts very rapidly to prevent extensive exposure of the astronauts to the trapped radiation there, but also were scheduled during periods when the sun was relatively quiet and no large flares were anticipated.

130

Protection against Radiation

All e-m radiation and especially its ionizing forms can be stopped by varying thicknesses of material or mass. X rays can be stopped by a few inches of lead. Alpha particles can be stopped by a sheet of paper. The intensity of gamma rays decreases rapidly as the total thickness of mass through which they must pass increases. These are the easy types of radiation to stop.

Cosmic rays are a different matter. Both primary and secondary cosmic rays are difficult to stop because of their high energies. When primaries plow into matter, they form secondaries that can be made up of very energetic charged particles as well as gamma and X rays.

Neutrons are difficult to stop because they're electrically neutral or have no charge. Some elements, such as beryllium, will absorb neutrons. Protection against neutrons depends on having mass available for neutrons to interact with and thus be stopped.

It was once thought that the best way to protect people against ionizing radiation in space was to build a pressure hull just thick enough to keep air in and ultraviolet radiation out.

Now we know that the presence of ionizing radiation in space is one of the biggest hazards other than the possible long-term biological consequences of weightlessness. Spacecraft may be able to function with the classical type of pressure hull referred to above, but habitats are going to have to have protection against not only the normal ionizing radiation that exists in space, but the enhanced radiation that comes from a solar flare.

But, as with the problem of weightlessness, there are several solutions to the problem of protection against ionizing radiation in space.

NASA astronaut Rhea Seddon assembles food items in a portable warmer. *(NASA)*

9 Nutrition and Sanitation

Nutrition and Sanitation on Earth

In common with an atmosphere and gravity, food is something you tend to take for granted unless you don't have any and become hungry. Food is usually something created by someone else and is obtained at a supermarket. The economic food chain is now so long and complex in high-technology cultures that most people have only the most rudimentary knowledge of food technology. You rarely think about food except in terms of dieting. In the United States and most Western nations, malnutrition and its associated diseases and syndromes are rarely seen even in the worst economic periods. On the other hand, overeating and its immediate and long-term consequences—indigestion and obesity—appear to be common denominators of the high-technology cultures of Earth.

A different attitude prevails when it comes to sanitation. This attitude is exemplified by the NASA euphemism for it, "waste management." In high-technology cultures, the normal physical acts of urinating and defecating aren't subjects for open discussion and, when circumstances demand their discussion, people use latinized words and euphemisms in place of their Anglo-Saxon or vernacular counterparts. While the physical act of eating can be a social occasion, waste elimination is a private action. Almost the opposite is true in other cultures of the world whose people are closer to the problems of raw survival. The difference between the two types of restrooms that have been

133

built for tourists at the Great Wall of China is an example of this; there's a "Western" bathroom with the usual ceramic equipment, and there's the native or Chinese bathroom with the series of holes in the floor for the Oriental "bombs away" technique. In the Western nations, sewage that includes human waste is treated to create a dried sludge which may be used directly for fertilizer or the privy is built right over the pigsty—because swine do an outstanding job of recycling human wastes into protein. In nearly all situations on Earth, we've solved the sanitation problem by putting human wastes where they'll be recycled back into the huge closed ecological life-support system of the planet in such a way that their presence and odor don't create disease or offend us.

Nutrition and Sanitation in Space

Nutrition and sanitation in space, however, take on new importance because, in essence, you're living in a very small Earthlike ecological system, whether you're in a spacecraft or a large Cole-type macro life colony. It's impossible to take food and waste for granted.

Food and sanitation technology in space resembles that in an oceangoing ship or a long-range aircraft. All the food for the journey must be put aboard at the dock or gate before departure. On a ship, waste can be dumped overboard into the huge ecological waste-recycling system of the ocean, but the waste collected in an airplane's lavatories is collected in "honey tanks" in the aircraft and transferred to a "honey wagon" at the destination airport, where the waste is put into an ordinary sewage-treatment system.

Because of the accumulated know-how of both nautical and aeronautical practice, we have a firm foundation upon which to build the nutrition and sanitation systems of spacecraft and space habitats. And because of studies of human nutrition, we already know how much food and what types of food must be supplied to a person in order to maintain good health.

Nutritional Requirements

As mentioned in an earlier chapter, you can be considered a heat engine that converts organic fuel into heat energy and motion, utilizing the oxygen of the air for the final step in the metabolic combustion process. Not all of this energy goes into producing useful work. You're only about 11 percent efficient. The remaining energy must be discharged into the surrounding environment either as heat or as physical waste.

Your metabolism is the sum of the processes by which your body uses nutrients and oxidants. Two major variables control individual metabolism, because each of us is different. Most of this difference is caused by the first variable, genetic inheritance, which determines the nature of the structure and function of your cells and thereby controls your individual ability to utilize the food, water, and air intake from the environment. The second variable is the sort of nutritional material you provide to your body. Two other minor variables are intermediary metabolism, which is the way your metabolic processes interact based on the chemical and physical properties of your environment; and the regulation of your cellular and organ interrelationships, especially by your endocrine system, which integrates your tissue activities in a way conducive to your continued health.

Although your energy requirements therefore vary from those of other people because of these variables and also because of differences in individual age, size, weight, state of health, and level of physical activity, you require between 1,800 and 3,600 kilocalories, or nutritional Calories, per day. Your minimum acceptable caloric intake seems to be about 1,600 Calories per day.

But these "normal" caloric intakes will depend upon your body mass, actual or desired. The normal daily caloric requirements for male adults is 21 Calories per pound of desired body weight, while the female adult requirement is 18 Calories per pound. Young, growing humans such as children and adolescents require as much as twice this normal daily caloric intake, depending upon their age.

However, for the next decade you probably won't have to concern yourselves with child nutritional problems in space because it's quite likely there will be no family life there until the first real space colonies become a reality—perhaps by the turn of the century. Since the space population will consist almost exclusively of grown, adult human beings, it's unlikely that many of the childhood nutritional and deficiency diseases will be encountered.

Water Requirements

Although you produce and exhale water vapor as one of the products of cellular combustion, you still require about two quarts of water per day as either drinking water or as part of your food. During periods of heavy physical activity with resulting profuse perspiration, or in high-temperature environments, this liquid intake may increase to as much as six quarts per day.

Second to a sufficient supply of oxygen at adequate partial pressure, your need for water is critical. People have survived without water for periods of up to seven days, but severe dehydration can begin within a few days, depending upon the individual and the environment.

The primary effect of dehydration is an imbalance in your body's acid-base relationship, reduction of intercellular fluids, and major changes in the electrolyte balance of the blood. All of these produce other symptoms that, depending upon the relative seriousness of the dehydration in a given individual, can create other disorders.

Depending on the design of the given life-support system in a spacecraft or space habitat, water may not be a problem. A good, operable water-recovery system properly maintained and sanitized will be able to provide more than enough clean, fresh, potable water. However, it must not be allowed to accumulate an excessive level of mineral impurities and it will have to be chemically treated to eliminate pathological organisms. This is old and well-known technology practiced daily by public-health and civic water-supply technicians.

When it comes to food, mere quantity or caloric content isn't enough. You require a "balanced" diet.

A typical balanced diet consists of 1,600 to 2,000 Calories of carbohydrates, 630 to 1,000 Calories of fat, 400 to 600 Calories of protein, and an adequate supply of both minerals and vitamins.

There's an enormous volume of information available regarding diets, foods, etc. There are all sorts of "special" diets. However, because of the nature of nutrition and sanitation during the next few decades of space living, you won't have to be a dietician to live comfortably in space. You should, however, be far more aware of nutritional requirements because, although you'll have a reasonably wide freedom of choice in foods in space, you won't have available the variety of foods that exists here on Earth.

You may well encounter the "micronutrient" deficiency diseases, which are caused by the lack of vitamins and minerals in a diet. Very small amounts of these micronutrients (0.00002 percent to 0.005 percent) are required. Most but not all native diets on Earth have evolved around locally available foodstuffs that possess adequate vitamin and mineral contents to satisfy human needs in this regard.

Vitamin-deficiency diseases are rarely encountered in the United States, Canada, and Western Europe these days because of the general abundance of most types of foodstuffs in those locales even in the depths of the harshest winters. Frozen vegetables, citrus juice concentrates, eggs, dehydrated milk, and other components of a balanced diet are shipped worldwide by air cargo. Thus, a breakfast in Fairbanks in December starts with orange juice and a business lunch in Manhattan in the same month starts with a crisp tossed salad. As a result, doctors rarely encounter vitamin-deficiency diseases even among the chronically poor, the transient population, or slum dwellers. This is true even in the most difficult economic times. Such was not always the case in the past. And, in some circumstances in space, this could again become the case if the logistics of food supply are interrupted.

The current listing of daily requirements of micronutrients is shown in Figure 9–1.

Micronutrient-Deficiency Diseases

You should be aware of the micronutrient-deficiency symptoms because some of them can be deadly in the space environment.

There are two basic types of vitamins: fat soluble and water soluble. Fat-soluble vitamins can be stored in your body, while an excess of water-soluble vitamins is eliminated from your body through urination or defecation.

Vitamin A (retinol) is a fat-soluble vitamin found mainly in fish-liver oils, liver, egg yolks, butter, cream, and green leafy and yellow vegetables. Vitamin A is readily and quickly destroyed by exposure to air and light. Most of your body's vitamin A is stored in your liver. Vitamin A deficiency primarily affects

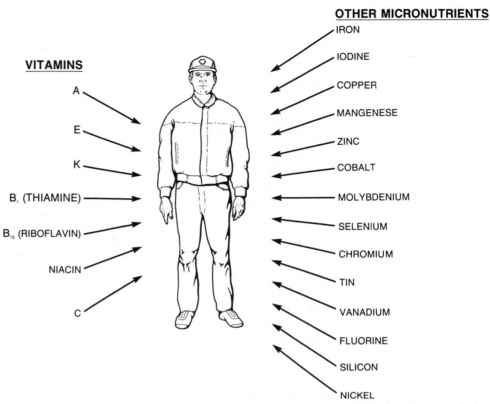

VITAMINS

A

E

K

B₁ (THIAMINE)

B₁₂ (RIBOFLAVIN)

NIACIN

C

OTHER MICRONUTRIENTS

IRON

IODINE

COPPER

MANGENESE

ZINC

COBALT

MOLYBDENIUM

SELENIUM

CHROMIUM

TIN

VANADIUM

FLUORINE

SILICON

NICKEL

Figure 9-1. The human body needs various vitamins and minerals, the so-called micronutrients, in order to remain healthy. *(Art by Sternbach)*

the eyes and leads to an incapacity of rod vision. Thus a lack of vitamin A causes night blindness, or an inability of your eye to adapt to low light levels. In its advanced stages, it can cause drying, thickening, or wrinkling of the conjunctiva, or white portion of your eyes.

Vitamin D, also fat soluble, is found in yeast, fish-liver oils, and egg yolks. It's also formed when you expose your skin to ultraviolet light. It's stored in your kidneys. Lack of vitamin D causes rickets, a disease that decalcifies bones and thus causes deformities in growing children. Although vitamin D deficiency wouldn't normally be a serious problem in space, its importance in calcium metabolism, where it has a major role in calcium transport in the intestines and bones, makes it critical because of the calcium-resorption syndrome of weightlessness. The importance of vitamin D levels in the human body with respect to the hypercalcemic hazards of weightlessness has not been evaluated yet.

The role of vitamin E (tocopherol) in human metabolism has not been thoroughly determined. This fat-soluble vitamin, present in vegetable oils and wheat germ, appears to be an intracellular antioxidant—that is, it seems to prevent or retard some of the aging processes in cells by maintaining the stability of cellular membranes.

Vitamin K, a fat-soluble compound present in green leafy vegetables,

137

appears to play a role in the ability of your blood to clot and prevent hemorrhage.

One of the most important water-soluble micronutrients is vitamin B_1, or thiamine. Although thiamine deficiency in its extreme form produces the disease called beriberi, in its less severe form it manifests itself in failure of appetite and certain forms of neuritis. Thiamine occurs abundantly in cereal grains and the husks of unboiled rice. Beriberi is primarily seen among people who subsist on rice whose hulls have been removed by milling; the husk contains most of the thiamine, but boiling the rice before husking disperses the vitamin throughout the grain, thus preventing its loss.

Vitamin B_2 is a water-soluble alcohol called riboflavin, which occurs in milk whey and egg white. Lack of riboflavin leads to lesions or open wounds of your skin and tongue.

The vitamin that isn't called a vitamin, niacin or nicotinic acid, is classed as a water-soluble B vitamin found in most protein-rich foods such as lean meat. Severe niacin deficiency causes pellagra, which usually occurs where corn is a major part of the local diet and where the corn has not been previously alkali-treated, as it is in Mexican tortillas. Pellagra appears initially as lesions of your skin, changes in your mucous membranes, burning of your mouth, abdominal discomfort, nausea, vomiting, and diarrhea. Pellagra can cause organic psychoses, excitement, depression, mania, delirium, and paranoia. Because of the hazards to space living that could be caused by pellagra, it's a micronutritional deficiency that should be watched for.

Homo sapiens is one of the few animal species that cannot synthesize vitamin C within the body. In comparison to other micronutrients, large amounts of vitamin C are required—70 milligrams or more per day. Citrus fruits as well as fresh vegetables and fruit are the best dietary sources of vitamin C. However, as a water-soluble vitamin, ascorbic acid is readily destroyed by oxidation and any excess intake of vitamin C is passed through your body quickly. Early voyagers across Earth's oceans ran into the consequences of vitamin C or ascorbic-acid deficiency: scurvy. This disease is characterized by weakness, lassitude, irritability, weight loss, little hemorrhages under the ends of the fingernails, and gross changes in the gums, which become swollen, purple, spongy, and bleed easily. Gangrene and loosening of the teeth may occur. On the other hand, vitamin C may act to retard various aging syndromes and has shown in some cases that it can prevent cancer as well as the common cold.

Noncritical Micronutritional Deficiencies

The symptoms and results of other micronutritional vitamin deficiencies aren't as critical or serious. Some of these micronutritional deficiencies can't be detected or diagnosed except by laboratory tests. Diagnosis and treatment can probably be left to the in-house medical department of a space habitat, as these can be detected during regular physical checkups. This is only one of many good reasons why regular medical checkups are extremely important in space.

Other micronutritional factors may be termed "trace elements." There are currently fourteen of these trace elements that are recognized as essential for warm-blooded animals and that occur in concentrations of less than 0.005 percent of your body weight. In order of demonstrated importance, they are iron, iodine, copper, manganese, zinc, cobalt, molybdenum, selenium, chromium, tin, vanadium, fluorine, silicon, and nickel. All of these are poorly absorbed and secreted by your body, and may therefore accumulate to the level of toxicity. Regular medical examinations should detect early signs and symptoms of toxicity, which, with regard to space living, appears to be far more serious than deficiencies of these trace elements.

In the new and artificial environments of space habitats, you may be far more susceptible to trace-element toxicities than you are on Earth, but only decades of experience and medical histories will provide the needed data.

With the exception of iron and iodine, it is generally uncommon for trace-element deficiencies to develop, no matter what diet you consume. Most deficiencies are caused by genetic factors and are manifested by metabolic disorders. Most of these will be detected in preliminary physical examinations on Earth, although you can anticipate that genetically caused trace-element deficiencies may begin to appear once family life becomes established in space habitats.

Space Food

An acceptable and varied diet is as important in space as it is in any other relatively isolated habitat. Not only should food possess the required caloric and micronutritional content, but it must also supply the necessary roughage because of your need for fiber in the diet.

Acceptability of food is just as important as balance. It's perfectly possible that sufficient food for a balanced diet may be available, yet you won't want to eat it.

To a large extent, food acceptability is primarily a cultural matter and secondarily a religious matter; taste and appearance, while factors, seem to be of much lower importance. People simply won't eat certain foods, no matter how hungry they become.

It may not be immediately obvious, but gas-producing foods should be eliminated from consideration in space menus. You may be working in modules with reduced atmospheric pressure or outside in a space suit at reduced pressure. Gas-producing foods can cause abdominal pains from gastric extension in addition to considerable flatulence, which is something that should be avoided as much as possible to prevent overloads of the air-system purification elements.

Food and Morale

Food is always a major morale factor. If the food isn't acceptable, your morale deteriorates and performance drops. You may even become nauseated and unable to eat what food there is. Variety seems to be a major acceptability

139

criterion, and any menu plan that duplicates in less than fourteen days becomes boring and therefore unacceptable.

Although all of this data has been around for several decades or more, very little attention has been devoted to it with regard to space habitation. Yet nobody will deny that food and nutrition are extremely important factors.

Space Food to Date

Food was taken into space on the initial manned flights, both American and Soviet. Major Yuri A. Gagarin, launched in *Vostok 1* on April 12, 1961, had 2,772 Calories of food available to him in pureed form stored in aluminum squeeze tubes. During his one-orbit space flight, Gagarin ate the contents of one tube of pureed food. On the first U.S. manned orbital flight, Colonel John H. Glenn had very primitive space food available to him: two economy-sized toothpaste tubes, one filled with applesauce and the other with beef stew. The use of squeeze tubes and straws allowed the cosmonauts and astronauts to handle food in weightlessness by squeezing it from the tubes right into their mouths.

As the length and complexity of space missions increased, a greater variety of food was made available. On the multi-man Soviet Voskhod flights in 1964, the cosmonauts had available such items as fruit juices and paste-type products in squeeze tubes and small bits of bread, marmalade, cheese, meat, confections, and fruit in plastic bags. The long-duration American two-man Gemini flights had a menu ranging from cereals, orange juice, and toast to meats, eggs, fish, and fruit salad, providing 2,500 Calories per day for each astronaut. The Gemini program also saw the first use of dehydrated foods in space. Solid foods were packed in bite-sized portions in tubes or in concentrated form enclosed in plastic bags, to which water was added. The Gemini menu, dehydrated,

---

Space Shuttle Menu Design

The Shuttle menu is designed to provide nutrition and energy requirements essential for good health and effective performance with safe, highly acceptable foods. In order to maintain good nutrition, the menu will provide at least the following quantities of each nutrient each day:

Nutrient	Unit	Amount	Nutrient	Unit	Amount
Protein	(g)	56	Vitamin B_{12}	(g)	3.0
Vitamin A	(iu)	5000	Calcium	(mg)	800
Vitamin D	(iu)	400	Phosphorous	(mg)	800
Vitamin E	(iu)	15	Iodine	(μg)	130
Ascorbic Acid	(mg)	45	Iron	(mg)	18
Folacin	(μg)	400	Magnesium	(mg)	350
Niacin	(mg)	18	Zinc	(mg)	15
Riboflavin	(mg)	1.6	Potassium	(meq)	70
Thiamine	(mg)	1.4	Sodium	(meq)	150
Vitamin B_6	(mg)	2.0			

Figure 9-2. The current NASA space-shuttle menu is designed to include all the vitamins, minerals, and other micronutrients required. *(NASA)*

Space Shuttle Food and Beverage List

Foods*

Applesauce (T)	Chicken and noodles (R)	Peach ambrosia (R)
Apricots, dried (IM)	Chicken and rice (R)	Peaches, dried (IM)
Asparagus (R)	Chili mac w/beef (R)	Peaches, (T)
Bananas (FD)	Cookies, pecan (NF)	Peanut butter
Beef almondine (R)	Cookies, shortbread (NF)	Pears (FD)
Beef, corned (I) (T)	Crackers, graham (NF)	Pears (T)
Beef and gravy (T)	Eggs, scrambled (R)	Peas w/butter sauce (R)
Beef, ground w/pickle sauce (T)	Food bar, almond crunch (NF)	Pineapple, crushed (T)
Beef jerky (IM)	Food bar, chocolate chip (NF)	Pudding, butterscotch (T)
Beef patty (R)	Food bar, granola (NF)	Pudding, chocolate (R) (T)
Beef, slices w/barbeque sauce (T)	Food bar, granola/raisin (NF)	Pudding, lemon (T)
Beef steak (I) (T)	Food bar, peanut butter/	Pudding, vanilla (R) (T)
Beef stroganoff w/noodles (R)	granola (NF)	Rice pilaf (R)
Bread, seedless rye (I) (NF)	Frankfurters (Vienna sausage) (T)	Salmon (T)
Broccoli au gratin (R)	Fruitcake (NF)	Sausage patty (R)
Breakfast roll (I) (NF)	Fruit cocktail (T)	Shrimp creole (R)
Candy, Life Savers, assorted	Green beans, french	Shrimp cocktail (R)
flavors (NF)	w/mushrooms (R)	Soup, cream of mushroom (R)
Cauliflower w/cheese (R)	Green beans and broccoli (R)	Spaghetti w/meatless sauce (R)
Cereal, bran flakes (R)	Ham (I) (T)	Strawberries (R)
Cereal, cornflakes (R)	Jam/Jelly (T)	Tomatoes, stewed (T)
Cereal, granola (R)	Macaroni and cheese (R)	Tuna (T)
Cereal, granola w/blueberries (R)	Meatballs w/barbeque sauce (T)	Turkey and gravy (T)
Cereal, granola w/raisins (R)	Nuts, almonds (NF)	Turkey, smoked/sliced (I) (T)
Chedder cheese spread (T)	Nuts, cashews (NF)	Turkey tetrazzini (R)
Chicken a la king (T)	Nuts, peanuts (NF)	Vegetables, mixed Italian (R)

Beverages

		### Condiments
Apple drink	Instant breakfast, vanilla	Barbeque sauce
Cocoa	Lemonade	Catsup
Coffee, black	Orange drink	Mustard
Coffee w/cream	Orange-grapefruit drink	Pepper
Coffee w/cream and sugar	Orange-pineapple drink	Salt
Coffee w/sugar	Strawberry drink	Hot pepper sauce
Grape drink	Tea	Mayonnaise
Grapefruit drink	Tea w/lemon and sugar	
Instant breakfast, chocolate	Tea w/sugar	
Instant breakfast, strawberry	Tropical punch	

*Abbreviations in parentheses indicate type of food T = thermostabilized, I = irradiated, IM = intermediate moisture, FD = freeze dried, R = rehydratable, and NF = natural form.

Figure 9-3. A list of the various types of food and beverages carried aboard the NASA Space Shuttle Orbiter, along with the ways in which they are packaged (NASA)

weighed about one pound per man per day. Gemini astronauts ate a pre-planned menu that included strawberries, sandwiches, bread, cocoa, pea soup, potato salad, chicken with gravy, and grape juice.

The food taken along on the *Apollo 11* flight to the Moon was almost a gourmet feast in comparison to the early orbital flights. The Apollo program saw the first use of spoon-bowl foods. The *Apollo 11* crew had a wide range of food items from which to select their daily menus. More than seventy different items of freeze-dried, rehydratable, wet-pack, and spoon-bowl foods were aboard. There were even snacks available.

On the Skylab missions, which lasted up to eighty-four days, all the food for

Typical Menu for the First Four Shuttle Flights*,†

DAY 1	DAY 2	DAY 3	DAY 4

Peaches (T)
Beef patty (R)
Scrambled eggs (R)
Bran flakes (R)
Cocoa (B)
Orange drink (B)
	Applesauce (T)
	Beef jerky (NF)
	Granola (R)
	Breakfast roll (I) (NF)
	Chocolate, instant
	  breakfast (B)
	Orange-grapefruit
	  drink (B)
		Dried peaches (IM)
		Sausage (R)
		Scrambled eggs (R)
		Cornflakes (R)
		Cocoa (B)
		Orange-pineapple
		  drink (B)
			Dried apricots (IM)
			Breakfast roll (I) (NF)
			Granola w/blueberries (R)
			Vanilla instant
			  breakfast (B)
			Grapefruit drink (B)

Frankfurters (T)
Turkey tetrazzini (R)
Bread (I) (NF)
Bananas (FD)
Almond crunch bar (NF)
Apple drink (B)
	Corned beef (T) (I)
	Asparagus (R)
	Bread (I) (NF)
	Pears (T)
	Peanuts (NF)
	Lemonade (B)
		Ham (T) (I)
		Cheese spread (T)
		Bread (I) (NF)
		Green beans and
		  broccoli (R)
		Crushed pineapple (T)
		Shortbread cookies (NF)
		Cashews (NF)
		Tea w/lemon and
		  sugar (B)
			Ground beef w/
			  pickle sauce (T)
			Noodles and chicken (R)
			Stewed tomatoes (T)
			Pears (FD)
			Almonds (NF)
			Strawberry drink (B)

Shrimp cocktail (R)
Beef steak (T) (I)
Rice pilaf (R)
Broccoli au gratin (R)
Fruit cocktail (T)
Butterscotch pudding (T)
Grape drink (B)
	Beef w/barbeque
	  sauce (T)
	Cauliflower w/cheese (R)
	Green beans w/
	  mushrooms (R)
	Lemon pudding (T)
	Pecan cookies (NF)
	Cocoa (B)
		Cream of mushroom
		  soup (R)
		Smoked turkey (T) (I)
		Mixed Italian
		  vegetables (R)
		Vanilla pudding (T) (R)
		Strawberries (R)
		Tropical punch (B)
			Tuna (T)
			Macaroni and cheese (R)
			Peas w/butter sauce (R)
			Peach ambrosia (R)
			Chocolate pudding (T) (R)
			Lemonade (B)

*Abbreviations in parentheses indicate type of food: T = thermostabilized, I = irradiated, IM = intermediate moisture, FD = freeze dried, R = rehydratable, NF = natural form, and B = beverage.
†Beginning with the fifth Shuttle flight, the menu cycle will be enlarged to six days.

Figure 9-4. Typical menus for a four-day NASA Space Shuttle mission. Compared to the "early days" of space exploration, space food is now almost a gourmet feast. *(NASA)*

the three manned missions was aboard the Skylab itself at launch. But because of the size of the Skylab habitat and the planned length of the missions, whole new approaches to food in space were taken. In retrospect, Skylab has become the watershed in terms of both space nutrition and sanitation. An actual wardroom was available where crewmen could choose their menus from a variety of frozen and dehydrated foods. Various kinds of meat, vegetables, cereals, and desserts were available. Foods that would stick to a spoon or fork were eaten with the usual table utensils, which were held in place on the table by magnets when not in use. Liquids were served in squeezable plastic containers. All three crewmen could eat together at the same time at the table, which contained facilities for heating or cooling food. Many of the modern convenience foods and food containers that we take for granted today on the supermarket shelves came about as a result of Skylab. For example, the pull-top can lid and the pop-top beer can are both a legacy of Skylab.

The Space Shuttle Orbiter has a galley capable of preparing meals for seven people in about twenty minutes' time. The Shuttle menus are similar to those for Skylab and contain many of the same sort of dehydrated, freeze-dried, wet-pack, and spoon-bowl foods that were aboard Skylab.

Growing Food in Space

It's quite likely that a plateau in space nutrition has been reached with the Space Shuttle. This is because all space food has a terrestrial origin. In many ways, a space habitat is like the living quarters at Prudhoe Bay, Alaska, or the various arctic and antarctic scientific research stations: they have to be supplied from the outside. People living in space during the remainder of this century will be almost totally dependent on terrestrial sources for their food supply.

Some pioneering experiments in astroagriculture have been made by Carl N. Hodges and his colleagues at the Environmental Research Laboratory of the University of Arizona. Plants have been grown without soil on the inner surface of a rotating drum. This simulated the environment of a rotating space habitat. However, until extremely large space habitats of the O'Neill and Cole varieties are built, and until the basic experiments of astroagriculture are completed in space, don't anticipate growing your food in space habitats.

However, growing your own food in space can eventually make space flight and living cheaper. A 1983 study by the Boeing Company for NASA showed that growing food in an Earth-orbit habitat occupied by as few as twelve people could result in significant savings if 50 percent of the food required was grown

Skylab pioneered the sort of food that's eaten in space today: freeze dried, foil packed, or put in pull-top cans. Skylab food was heated in this modular tray. *(NASA)*

143

The Space Shuttle Orbiter food galley is compact and a model for things to come in space. *(NASA)*

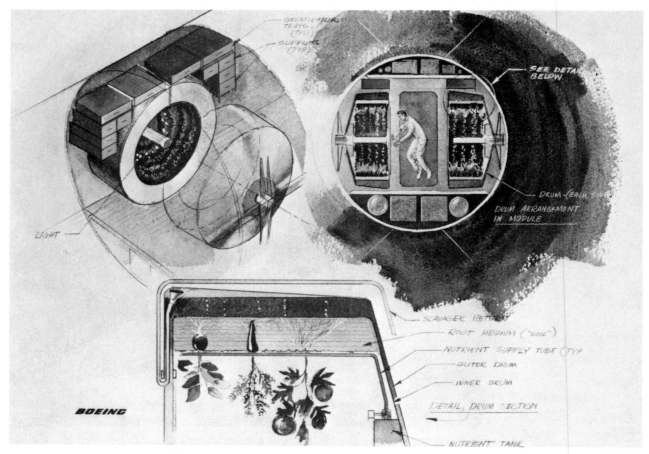

Figure 9-5. Food crops can be grown in the weightlessness of space. The Boeing Company and the University of Arizona have studied and grown plants in rotating drums that provide pseudogravity needed. *(Boeing Company)*

aboard. The savings acquired by on-board food production, however, are dependent upon time. Short-duration space flights and short-term habitats that are expected to operate for less than about five years don't show any savings if food is grown aboard. As a longer operational life is anticipated for a habitat and as a greater percentage of food is grown aboard, greater savings are produced. The Boeing study showed that, over a fifteen-year time period, growing 97 percent of the necessary food in an Earth-orbit habitat for twelve people could save at least $68 million.

Therefore, some food will come from the early closed-cycle life-support systems where broad-leafed plants such as the pumpkin are used to convert carbon dioxide to oxygen by photosynthesis. Small amounts of food may also come from little individual hothouse gardens in your personal quarters. But it will be many years before you'll live in the fully balanced ecological systems of the O'Neill and Cole colonies, where 100 percent of your food will be grown. In the meantime, your steak and cake will have to come from Earth until we reach that level of know-how and subsistence in space. You'll be tied to Earth by three umbilical cords: nitrogen, lithium hydroxide, and food.

The logistics of supplying all space food from Earth requires a great deal of energy in the form of rocket propellant. Weight is still a prime factor in space travel. Approximately 8.3 pounds of food and water per person per day are required. However, most food contains a very large percentage of water—which weighs about eight pounds per gallon. And there's abundant water available in space habitats from human metabolic processes and respiration. The prevailing logic is: Why lift water? Use dehydrated, partially dewatered, condensed, or paste foods.

There will come a day, however, when the first meal is cooked using food that's been grown in space. This is likely to take place sometime early in the twenty-first century.

Food Preservatives and Preservation

In the meantime, since food will have to be supplied to habitats, this brings up the obvious problem of food preservatives. Fresh food is going to be extremely rare in space during the early years. Because of preparation and transportation times needed to get food to space habitats, nearly everything eaten in space will be preserved.

The art and technology of food preservation is centuries old. Meat was preserved by drying (jerky), soaking in concentrated saline solution (salted foods), and by smoking. Brine, spices, and dehydrating have also been used.

Today, nitrites and nitrates of sodium and potassium are widely used to cure meats and other foods because these chemicals inhibit the growth of microbes, especially *Clostridium botulinum*, found naturally in lakes and soils. Nitrates become nitrites and are therefore used in foods that are stored for long periods of time; nitrates are also converted to nitrites by the chemical action of saliva.

The use of nitrites and other food preservatives is being questioned today because some studies have indicated that nitrous acid created from nitrites by the stomach's own acid environment can cause changes in DNA that might lead

145

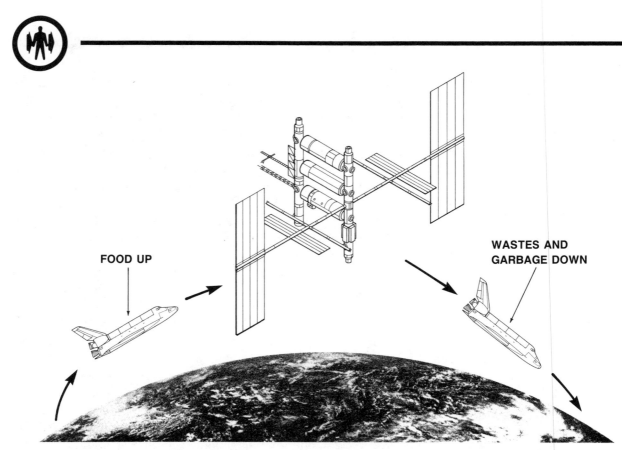

Figure 9-6. For near-Earth habitats, food and other consumables must be brought up from Earth for use in space. Wastes are taken back to Earth for disposal. *(Art by Sternbach)*

to cancer and birth defects. Furthermore, other studies have shown that nitrites can combine with the amine compounds normally found in people as well as in some foods to form nitrosamines, which are potent carcinogens. But many foods naturally contain chemicals that we *know* are hazardous or poisonous. For example, shrimp contain a high level of arsenic. Onions are full of highly complex carcinogenic-like chemicals. Cauliflower contains a poisonous thiocyanate.

Food additives, whether they be for preservation or cosmetic purposes, aren't necessarily bad chemicals. All the data isn't in yet, and one is therefore forced to choose between a slight risk of cancer or a very high probability of food poisoning or botulism. Thus, the risk-benefit ratio must be kept primarily in mind. In spite of some scary laboratory reports on food additives, people today are living longer, healthier lives and eating balanced diets having a great deal of variety and choice.

The Basic Sanitation Problem

The other side of the coin from nutrition is sanitation. Human beings don't utilize a large percentage of the food volume they eat. Some of it is used by the

146

gastrointestinal tract as roughage and passed through. The waste-management system of the human body also uses the liver, kidneys, and intestinal tract as means for getting rid of wastes and poisons that result from metabolic processes. Although a small portion of this waste elimination takes place through perspiration, most of it is eliminated from the human body through urine and feces.

Any vehicle or habitat, regardless of where used or located, must have some means of human waste management by a sanitary method.

For automobiles, it's simple: the roadside rest stop or gas-station rest room or, in an emergency, a bush.

Railroad trains simply dump everything overboard along the right-of-way.

Boats and ships also dump everything overboard into the water.

But in the air or in space, you can't pull over to the side of the road for a bush stop. You shouldn't or can't dump the wastes overboard. There isn't any easy way to dispose of urine or feces.

Early aviation—and even some small general-aviation aircraft today—had provisions for handling urine by methods as simple as a coffee can under the seat or an overboard relief tube. But there's no provision for defecation in such small craft even today. (I know; I have a four-place airplane with a coffee can under the front seat.)

As aircraft got larger and began to carry more people for longer periods of time, on-board toilets were developed. The first of these was for the Douglas DC-3 in 1934. This amounted to an airborne privy. The Boeing 707 jet airliner was the first to utilize a flush toilet. The Boeing 747 has twelve of them.

The History of Space Sanitation

Space sanitation has followed a similar developmental path.

Project Mercury astronauts could urinate into collector bags built into their space suits but had no provisions for defecation. They were fed low-residue diets prior to their flights, the longest of which was a little over thirty-four hours in duration—so there was no need to consider defecation provisions in the limited volume and weight situations of the Mercury space capsule.

Project Gemini, however, with its planned fourteen-day mission duration, was another matter. Astronauts could not be fed low-residue diets that would eliminate the need to defecate for two weeks. Therefore, a piece of sanitation equipment was developed that became a regular part of space travel and habitation: the Gemini Bag.

The Gemini Bag is a clear plastic container. It contains a glove into which you place your hand so that you can attach the adhesive-ringed neck of the Bag to your anus. After defecation, you remove the Bag and immediately seal it. You must then knead the bag to break the seal on an inner container of germicide that, when mixed with the fecal matter, prevents bacterial growth and gas formation. The Bag is then disposed of in the spacecraft or habitat trash containers.

147

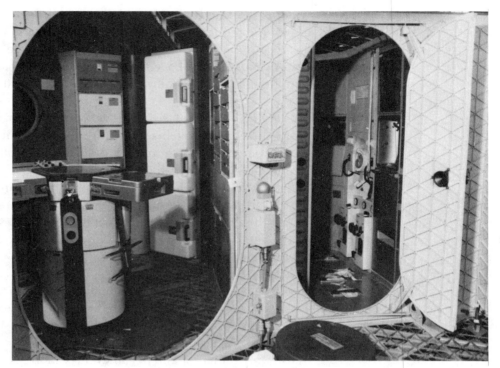

The Skylab wardroom/kitchen (at left) and the bathroom (right) have served as the model for all food and waste-management systems in space. *(NASA)*

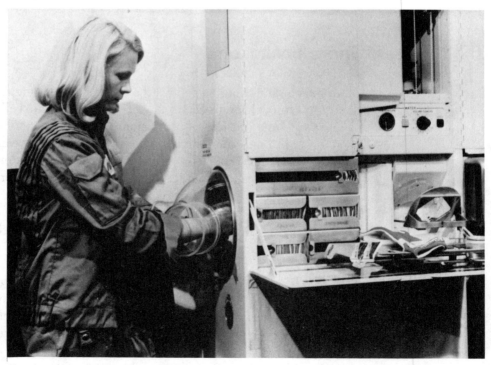

Personal hygiene in weightlessness means keeping liquids under control and using lots of absorbent wiping cloths. *(NASA)*

The Gemini Bag wasn't and isn't the real answer to space sanitation, as all of the Gemini and Apollo astronauts learned. Former astronaut Ron Evans points out, "There ain't no graceful way." But the Gemini Bag concept is useful for emergencies.

Modern Space-Sanitation Systems

Skylab saw the first space toilet, and the Space Shuttle Orbiters have a design that's a great improvement over the Skylab unit. The Shuttle toilet can be used by both men and women. It can also handle the disposal of urine, feces, toilet tissue, and "whoopee sacks" (or, as NASA prefers to call them, "emesis collectors"). It can even handle the disposal of Gemini Bags previously used in emergencies. It has a seat to which you strap yourself if necessary. A flow of air around the base of the seat replaces gravity and draws fecal matter into the commode, where motor-driven slinger tines shred it and deposit it in thin layers on the commode walls. The air then passes through a debris filter, a hydrophobic filter, and fan separators. The commode is serviced after the Shuttle returns to Earth at the end of its mission. The Space Shuttle toilet is the spacegoing version of the jet airliner commode.

In early spacecraft and habitats such as Gemini, Apollo, and Skylab, urine was disposed of by dumping it overboard into space, where it immediately crystallized. Gemini Bags were stored for disposal on return to Earth. In Skylab, the fecal-collection bags used in the toilet were put away with the trash in the waste-collection tank of Skylab, where it eventually burned up when Skylab entered the Earth's atmosphere and was destroyed over Australia. The Space Shuttle Orbiter retains all waste for postflight disposal.

The Shuttle method of on-board retention of both feces and urine is the technique that will be used henceforth, because something such as a urine dump actually contaminates the space environment around the spacecraft. This in turn may greatly affect experiments, sensors, and industrial processes that depend upon a clean, uncontaminated vacuum around the craft or habitat.

It's quite likely that space sanitation has also reached a plateau of development, too. Just as food will have to be brought up from Earth, so waste will have to be returned to Earth in the foreseeable future.

Closed-Cycle Systems

Although systems to recycle urine into potable water were demonstrated in the laboratory as long ago as 1959, no closed ecological systems have yet been used in space. They won't make their appearance until the economics of space habitation force their development and use. The open-cycle logistical nutrition and sanitation techniques are now well known, tested, and dependable. These are important factors, because reliability is something you'd better have in all space systems. There's too much we don't know.

149

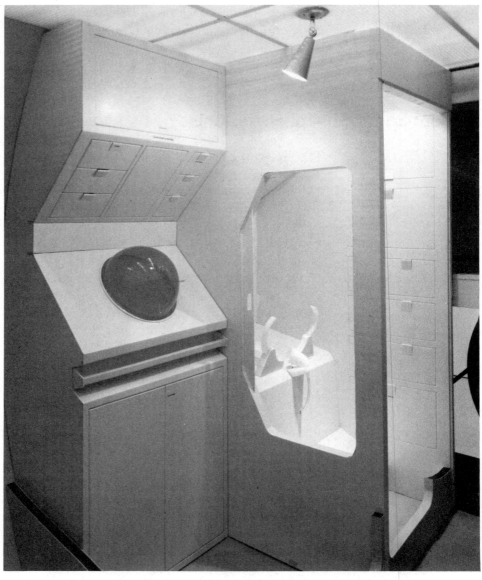

Future space habitats will have personal-hygiene equipment and waste-management systems that resemble these in mockup by Boeing. *(Boeing Company)*

A closed-cycle system can be designed. But maintaining balance in a closed-cycle system is something we're going to have to learn how to do in space habitats.

The classical open-cycle nutrition and sanitation systems that have been used in space to date are feasible and economical even for space habitats housing up to a hundred people. However, there's a crossover line between the technology and economics of open systems versus closed systems somewhere beyond the hundred-person space habitat in the Earth-Moon system. When

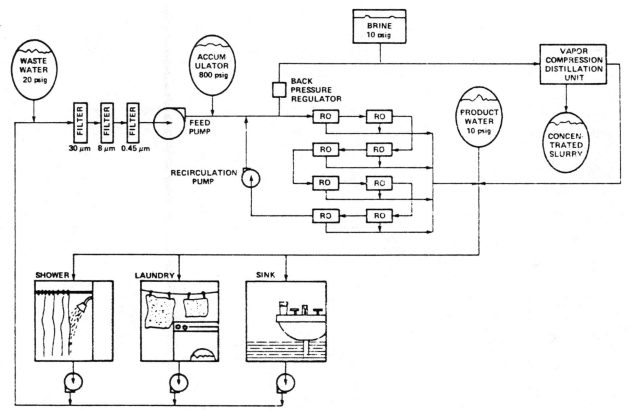

Figure 9-7. A generalized block diagram of the water portion of a closed ecological life-support system. Such systems will be required for long-term space missions or in habitats where supplying consumables from Earth is expensive or difficult. *(NASA)*

we talk about lunar outposts or operations beyond the Earth-Moon system, it's implied that closed systems will be used. But at this time, such advanced life-support systems are merely conceptual or, at best, laboratory demonstrations. Once we have reliable closed-cycle life-support systems capable of recycling carbon dioxide to oxygen, urine to water, and feces to food, we'll be free of most ties to Earth except for the need to obtain nitrogen there. But that's twenty-first-century technology.

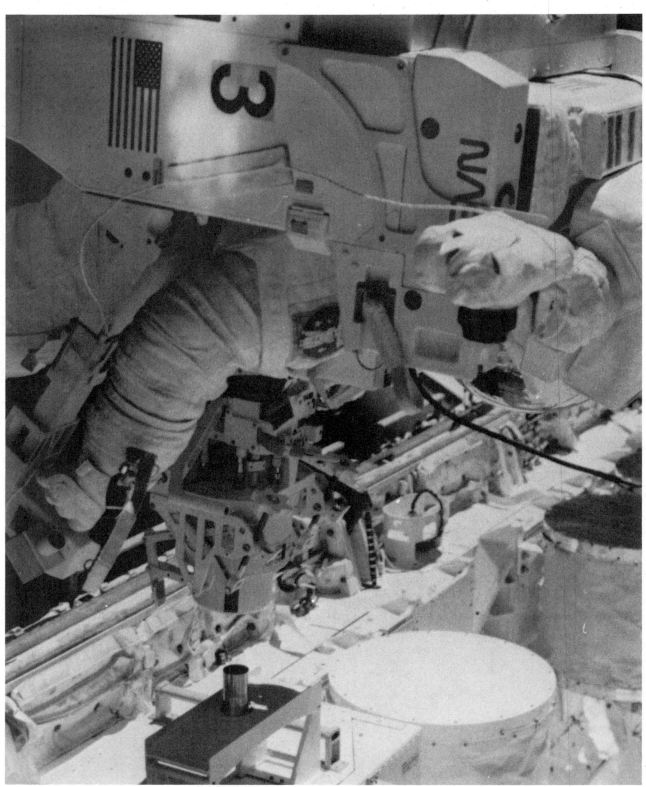

Working in space *(NASA)*

10 Working in Space

A human being is the best all-around general tool user ever evolved on Earth. As discussed earlier, the human body commanded by the human brain is extremely versatile and far outstrips the abilities of other animals, including other primates, on Earth.

The Basic Problem of Working in Space

You should have no trouble working in space if there's pseudogravity created by spinning a part of a space habitat (although there are other problems involved when you work in a rotating habitat). But when it comes to working in weightlessness, astronauts and cosmonauts both have encountered unforeseen difficulties, and you'll run into the same ones.

These difficulties shouldn't have been unforeseen because to some degree they can be and are encountered on Earth in specific environments.

You can experiment yourself to see what the major problem with weightless working amounts to.

With a snorkel or scuba gear, get under water. This is the closest you can be to weightlessness on Earth except for brief periods of weightlessness that can be achieved in aircraft flight.

Many swimming pools today are equipped with one or more underwater lights to illuminate the pool for after-dark swimming. Go under water with a

153

Even the thermal tiles on a space shuttle can be repaired in space. (NASA)

screwdriver or other tool, try to remove the lens from the underwater light, and then try to replace the light bulb.

When you twist the screwdriver counterclockwise to loosen a screw, your body will twist clockwise. You'll try to grab something to hold onto and brace yourself against so your body won't twist in the opposite direction to the action of your arm, hand, and fingers. Even if there's a nearby handhold to grasp, you'll discover that twisting that screwdriver or wrench, or trying to screw the light bulb out of its socket, is an extremely difficult task. Furthermore, it takes a lot of hard work because you can't brace your body and therefore you're actually doing twice as much work as normal. The extra effort goes into trying to keep your body from twisting and turning in reaction to the work you're trying to do.

The Gemini EVA Experience

The above phenomenon should have been fully anticipated for human activities in weightlessness, but it wasn't. It showed up on the second U.S. space walk, or EVA, during the flight of *Gemini 9* on June 3, 1966, when astronaut Gene Cernan went outside the space capsule. The first Gemini EVA with Edward White on *Gemini 4* a few months previous had revealed no problems with a human being moving freely around in space outside the spacecraft. But Cernan discovered that it was exceedingly difficult to move along the side of the Gemini spacecraft to get to the rear end, where a bulky Astronaut Maneuvering Unit (AMU) was stored and waiting for him. In trying to move along without handholds, Cernan had to rest repeatedly. His space suit's faceplate fogged over inside with moisture from his own perspiration that the space-suit system couldn't handle. By the time he got back to the AMU, the life-support system of his suit was overworked, and the visor fogging got even worse. Then Cernan had trouble extending the folding arms of the AMU because he didn't have any way to brace himself. The visor fogging got so bad that the mission commander, astronaut Tom Stafford, ordered Cernan back into the capsule after about two hours.

Wiping away the fogging of a space-suit visor falls into the same category as having to scratch your nose: In a space suit, you can't do it. There was no way that Cernan could wipe the condensation off the inside of his visor.

Working in weightlessness had proven itself to be far more strenuous than anyone had anticipated. This was amply confirmed during the *Gemini 10* and *Gemini 11* flights, during which EVAs had to be cut short because the astronauts exerted themselves beyond the capabilities of their life-support backpack units. The problem was partly solved on *Gemini 12*, the last Gemini flight, when Edwin Aldrin used body tethers, foot restraints, and handholds to brace himself while performing work in the weightless environment.

But something had to be done to discover exactly what was causing astronauts to become overexerted, how tools and other equipment could be properly designed for best operation in weightlessness, and how to train astronauts to work in weightlessness before they were expensively placed in orbit.

NASA trains astronauts to work in weightlessness by using a huge underwater facility, a "neutral buoyancy simulator," which produces the closest analog to weightlessness. *(NASA)*

155

NASA astronaut Shannon Lucid works in the NASA Marshall Space Center's neutral-buoyancy water tank learning to use a power tool in simulated weightlessness. Normal distortion makes her head look small in the transparent helmet. *(NASA)*

Neutral Buoyancy EVA Simulation

Someone—I don't know who came up with the idea, probably someone who had tried to replace the light in his swimming pool—suggested that astronaut EVA training could be conducted under water with the astronaut in a space suit and ballasted to achieve "neutral buoyancy" so that he wouldn't sink or rise.

So the world's biggest indoor swimming pool was built at George C. Marshall Space Flight Center in Huntsville, Alabama. It's a 1,300,000-gallon water tank 75 feet in diameter and 40 feet deep. Full-scale spacecraft and space habitats can be placed in it, and space-suited astronauts using special life-support backpacks and weighted to achieve neutral buoyancy can work assisted by scuba divers in the closest thing to weightlessness that can be achieved in the persistent gravity of Planet Earth.

Neutral buoyancy simulation worked, and it will probably be used for space-worker ground school during the years to come. Flying airplanes in weightless arcs through the sky can create perhaps 30 to 40 seconds of actual zero-g, but neutral buoyancy immersion permits the nearest equivalent to weightlessness to be experienced for as long as the air in the suit backpack holds out.

Most of the techniques of working in weightlessness were subsequently worked out in the neutral buoyancy simulators and then tested in the space environment itself.

Working in Weightlessness

To work best in weightlessness, you must use foot restraints. These will hold your feet firmly to the "floor" as gravity does on Earth and will permit you to use your hands and arms normally. Your body already possesses the necessary muscles and nervous-system programming to overcome the reaction forces involved while using tools and otherwise working in space if you can keep your feet in place.

It's also possible to work quite well if you can wedge your body between spacecraft or space-habitat walls or equipment in such a way that your actions don't rotate your body. Friction still works in weightlessness, and most astronauts thus far have made just as much use of this fact as they have of the special foot restraints of their spacecraft.

These are basic techniques used by underwater free divers to use tools and otherwise work on submerged equipment. Although whole-body restraint is workable, most divers manage to wedge themselves between rocks or between pieces of equipment in order to push, pull, or twist. In cases where it's possible for them to wedge their feet in place, they've found that they can perform most push-pull-twist tasks.

A foot restraint is the most important accessory for a person working in weightlessness because it allows the feet to be firmly planted. *(NASA)*

Figure 10-1. Working in a foundry in space utilizing concentrated solar energy *(Art by Sternbach)*

Special Tools

There will be situations in weightlessness where it won't be possible to provide restraints to prevent your body from moving when you try to do something. Although ordinary tools such as wrenches, screwdrivers, cutters, saws, and welders can and will be used in space, there will be special zero-g tools for those applications where ordinary tools won't work or where you can't find a restraint for your feet or for your body.

These tools are already in existence, although they've never been used in space. Back in 1965, the problems of push-pull-twist tool operation in zero-g were anticipated by engineers. The U.S. Air Force let a contract with the Martin Company to develop various zero-g tools. Some of these were to be flown on *Gemini 8* as Defense Department Experiment D-16. As already pointed out, Eugene Cernan got into trouble in the *Gemini 8* EVA and was unable to test these tools, which were located in the EVA maneuvering unit on the aft end of the spacecraft. It was rescheduled for *Gemini 11* but apparently wasn't tested on that flight because Dick Gordon ran into problems with temperature overload of his space suit and weightlessness-induced fatigue.

Among the zero-g tools already developed are:

A zero-reaction power tool.   This unit is something like the ⅜-inch portable hand drill found in most earthside home workshops. It looks something like an ordinary power drill but doesn't possess the twisting reaction force of a standard power tool. It operates something like a very fast impact wrench of the sort used by service-station mechanics to remove and install lug nuts when changing a tire on a car. Designed and built by Black & Decker Manufacturing Company, this zero-reaction power tool reduces the reactive torque to $\frac{1}{10,000}$ of its output torque through an arrangement of bearings, springs, gears, a counter-rotating barrel, and a repetition rate of 1,800 impacts per second. To get some idea of the difference between this space tool and an ordinary impact wrench: the service-station tool has a reactive force of $\frac{1}{50}$ of its output and operates at perhaps 10 impacts per second. In space, this would be enough to spin you like a top if you didn't use foot restraints to hold your body in position. The zero-reaction power tool uses rechargeable nickel-cadmium batteries and would accept into its chuck various screwdriver and socket accessories.

A ratchet screwdriver. This is something like the currently available screwdriver with the ball-shaped handle and ratchet action. It was developed because an ordinary screwdriver is extremely difficult to grasp when you're wearing a pressurized space-suit glove, whereas the ball-handle is easy to grip.

A spring-driven hammer. This hammer has a pistol grip something like an ordinary rivet-driving gun except that it operates at a much higher hammering frequency.

An aluminum fingernail. This fits over the finger of a space-suit glove so that you can pick up small parts. To get some idea of this problem, try to work while wearing a pair of very heavy, stiff gloves. Even on Earth, this metal fingernail would be useful.

Velcro and adhesive buttons. These are placed at critical places on the spacecraft so that tools equipped with compatible adhesive strips could be temporarily placed out of the way while working. You can't just lay down a tool in weightlessness; it will float there or, most probably, float away. (Velcro strips are already in wide use aboard the Space Shuttle Orbiter for just this purpose.)

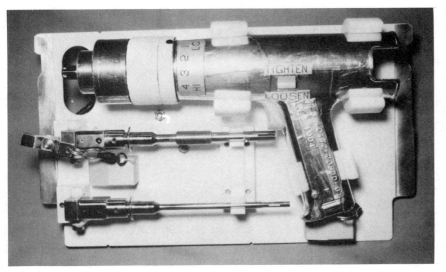

Space tools used in Space Shuttle missions (NASA)

Electric lights. Affixed to both sides of the space-suit helmet, they illuminate the work area. They're even handy on Earth, and they're used by people other than doctors. There are several head-mounted lamps available in hardware stores and handyman shops.

A special chain with an internal cable. When the cable is pulled tight, the chain is locked into the position it's in at the time. This was developed in the late 1960s at General Electric Missile and Space Division in Valley Forge, Pennsylvania. You can make one of these yourself to test it out. Get an ordinary wooden or metal set of children's play beads. Remove the internal string and replace it with a flexible cable. At one end, attach the cable securely to the end bead. At the other end, install a pistol grip that will pull tightly on the cable. When the cable is loose, the bead chain is completely flexible and can be bent and twisted at will. However, once the cable is tightened, the bead chain locks itself in precisely the position it is in at the moment you tighten the cable. The bead chain may be twisted like a pretzel, yet it will lock up firmly in that shape when the cable is tightened.

A zero-g tool and parts holder. This consists of a series of small metallic or bristle brushes mounted bristles-up on a plate. Parts with holes such as nuts or

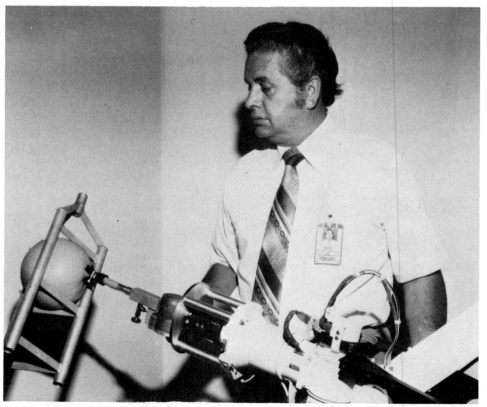

An example of the simplicity of weightless tool concepts. Keith Clark used a soda straw and a toy balloon to work out the details of this simple device, which will be able to grasp and transport aluminum girders in weightlessness. *(NASA)*

small cover plates can be pushed down on the brushes to hold them in place. Or tools can be wedged in between brushes.

Zero-g tool design for space workers is going to be a fun field to get into, and the need for this sort of thing is just around the corner. It's totally in its infancy, and there are undoubtedly hundreds of special space tools just waiting to be invented.

However, unless you're willing to test them under water, you may have to wait until you get into space to do your inventing. We can anticipate many of the factors involved in working in weightlessness, but not all of them, because we have a distorted viewpoint. All of us at this time were born and raised on Planet Earth, and gravity is a commonplace thing to us. We really don't fully understand everything about living and working in zero-g yet. We can do some interesting guessing, but you will have to live and work in space itself to learn what you really need to invent in terms of new tools or modifications of existing tools.

Physiological Aspects of Space Work

As we discussed in an earlier chapter, your metabolic heat output differs greatly between your resting state and your active working state. During sleep, your energy expenditure is about 65 Calories (Cal) per hour. At rest lying down, it rises to 80 Cal/hr, and, when sitting up, to 100 Cal/hr. During light exercise, this jumps to as much as 200 Cal/hr. and, during heavy physical work, to 500 Cal/hr. Under normal conditions, you're 11 percent efficient as a heat engine. This means that 89 percent of your metabolic heat output shows up as increased perspiration rate, increased respiration rate, and slightly elevated body temperature.

Life-support systems will be built to handle your anticipated additional heat loads when you're working in a reasonably large space vehicle or habitat. And the life-support system will be designed with adequate reserve capacity to handle periods of heavy physical work where you're dumping as much as 425 Cal/hr. into the environment, heat that must be removed from the environment in order to maintain comfortable temperatures and humidities.

Design of Working Space Suits

A space suit—which should be considered as a small, body-fitting, personal, soft spacecraft built for one—can pose a different problem. Because of its greatly reduced internal volume and the limitations of bulk placed upon any self-contained life-support backpack, a space suit is extremely sensitive to your metabolic heat output as well as to external environmental stresses such as heat and other radiation from the sun. This situation may be less critical for suits that don't have their own life-support backpacks and are instead linked by umbilical lines to the larger life-support systems of spacecraft or habitats. But early umbilical-linked space suits used in the Gemini program were exceedingly

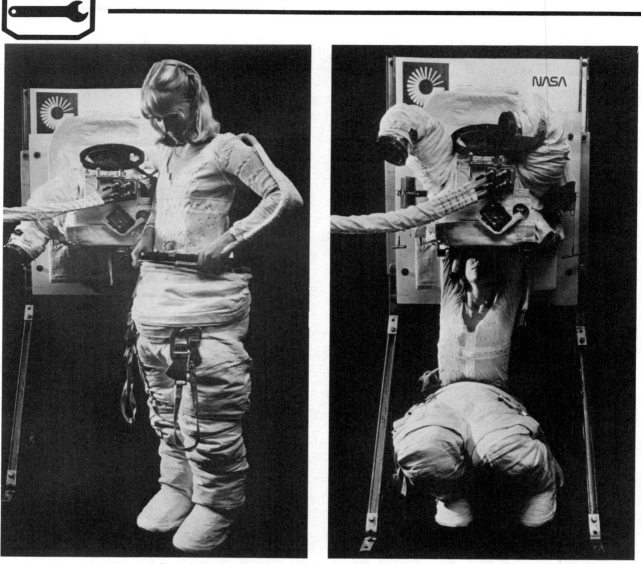

Sequence showing Hamilton Standard test engineer Jocelyn Johnson donning the Space Shuttle space suit. Liquid cooling garment, lower torso, and hard upper torso are shown in first photo. *(Hamilton Standard)*

limited in their abilities to handle human heat output overloads. Later space suits have been improved in this regard.

Space suits used in the NASA Space Shuttle are vastly better in this regard than the primitive Gemini ones. But a prospective astroworker must still receive underwater training in a neutral buoyancy tank in order to work effectively in space.

There's still a broad technical gap between today's space suits and suits that might be used by space workers with little or no training and in work environments of widely different characteristics.

The ultimate space suit will not only come in standard sizes—small, medium, large, and extra large, like clothing—but will also have been developed into a highly reliable system with the capability to safely handle a

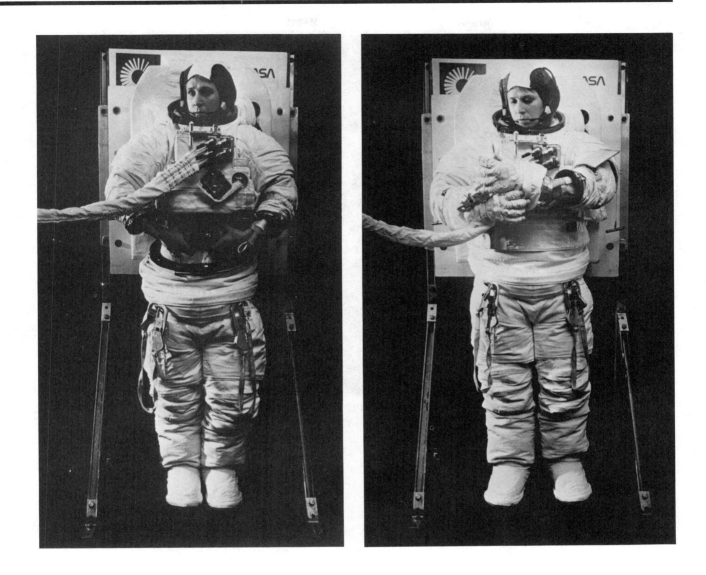

human heat output of 500 Cal/hr. for at least an hour (about the length of time an average person can work at a heat-expenditure rate of 500 Cal/hr).

The life-support system will probably have to be contained in a backpack, because it can't be distributed around the space suit and because the backpack location interferes least with your movements and working activities. Such a system will be full of redundant equipment so that the failure of a single fan, oxygen regulator, or other critical component won't create a major crisis for you.

The space suit will be easy to get into and out of. In this regard, the NASA Space Shuttle suit is a big step forward: it can be donned in a few minutes. Ease of donning and doffing may become a critical element in case of an emergency—and there will indeed be emergencies.

163

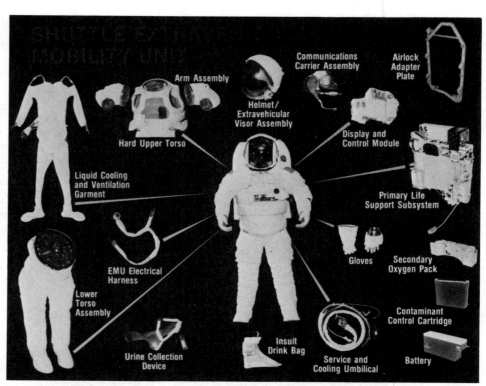

Figure 10-2. The parts of the most advanced current space suit, the unit used in the Space Shuttle Orbiter *(NASA)*

Teleoperators

However, developments in the new field of *teleoperators* may eliminate the requirement for humans to "suit up" and work in space suits in space itself.

A teleoperator is just what the word describes: an operating machine that's directed from a remote location by a human being.

A crude example of a teleoperated device is the power steering of an automobile. You don't actually turn the front wheels of a car by hand. You do it through an incredible collection of shafts, levers, joints, bearings, and a "strength booster" system called power steering. The system is designed to be redundant so that you can still steer the car if the booster fails. The assist unit is a hydraulic system that uses a pressurized liquid provided from an engine-driven hydraulic pump. When you turn the steering wheel, this signals the hydraulic system to push or pull on the mechanical system that turns the front wheels.

You could actually be seated anywhere in the car provided that the steering wheel—or steering command lever—was available to you and connected by hydraulic hoses to the steering system itself. In a like manner, you could be totally removed from the vehicle and still activate the steering system remotely by means of a radio link from you and a steering wheel seated anywhere external to the car. Such systems are often used during automotive crash experiments.

164

The ordinary radio-controlled model airplane is one form of teleoperator, although such a device is usually operated for recreational purposes. However, if the radio-controlled model airplane carries a television camera to look at the ground, it becomes a crude teleoperator. Target drone aircraft for use with antiaircraft missile testing or tactical fighter gunnery training are also teleoperators.

Many engineers and technologists believe teleoperators will find extensive use in space work. You can be seated comfortably inside the spacecraft or habitat and—by means of a radio control "up link" to the machine and an activity-reporting "down link" from the machine—perform most of the activities with such a teleoperator as if you were on the spot.

Teleoperators are basically a new field of endeavor, although the old Surveyor lunar lander spacecraft was a teleoperator that dug into the surface of the Moon under human command as long ago as 1967. So was the Viking Mars lander of 1976.

The Teleoperator Problems

At the moment, there are several basic problems with teleoperators.

The first of these has to do with a subject discussed earlier: how to build a machine that will do everything that a human being can do. It's totally possible

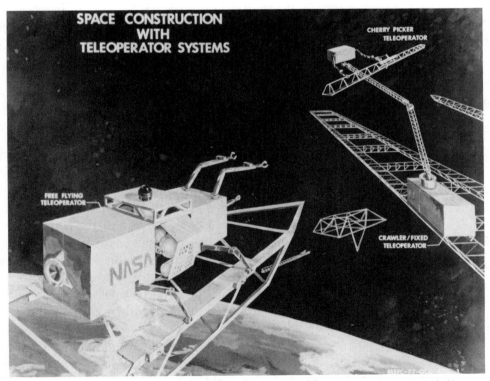

Figure 10-3. Teleoperators are robots that are remotely controlled on a real-time basis by people. They can be used in places and for work that are hazardous for humans. *(NASA)*

165

to build a machine that will duplicate the human arm-wrist-hand-finger actions; but it will be big, clumsy, complex, and nowhere near as neat as its human counterpart. Therefore, until the technology of teleoperators progresses significantly, teleoperators will either be somewhat larger than human beings if teleoperator universality is required, or the teleoperator will be designed and built to do specific tasks.

A teleoperator intended to perform specific tasks will be quite restricted in its utility. It will be able to do only the job it's designed to do. If it becomes necessary for it to do something else, you as the operator must be extremely clever in order to manipulate it in a manner for which it may not have been designed.

The biggest teleoperator problem is command circuit-time delay. If the teleoperator uses a radio up-link and down-link, the problem of time delay is inconsequential if the teleoperator is close to you—say, within 50,000 miles. At distances greater than this, the performance of the teleoperation system— teleoperator and human operator—become affected by the fact that radio signals travel at a finite speed of 186,000 miles per second.

It's just barely possible for you to direct a teleoperator on the Moon at a distance of some 240,000 miles. There's a time delay in the system of as much as four seconds. This is due to the time required for the radio signal to get from you on Earth to the teleoperator on the Moon, plus the time required for the down-link signal to report from the lunar teleoperator back to you that the command or action has been received and is being carried out.

If you're a good human teleoperator director, you can learn to handle this four-second time delay. But when it comes to handling a teleoperator on the surface of the planet Mars, there may be a time delay in excess of an hour to contend with.

Command Override Robots

Teleoperator systems with long time delays can also be considered to be "command override robots." In a teleoperator system with a long time delay, you can't directly control the remote unit but must give commands to the unit's computer; the computer then controls the remote unit while at the same time reporting back its compliance and the actions of the unit. Thus the system becomes one where you issue commands to a robot, which then carries out a complex series of activities without direct human action. This is quite different from an ordinary teleoperator system where the remote unit follows human commands on an action-by-action basis with the subsequent human commands depending upon the nature of the information sent back by the remote unit on the down-link.

Such a command override robot must have a complex computer and computer software. To some extent, it must also be self-aware and self-programming so that it doesn't get itself into trouble before it can call for help and receive human instructions on how to proceed to get out of trouble.

166

Figure 10-4. When time delays in the command links become very long, such as between Earth and Mars, robot explorers will have to be equipped with "artificial intelligence" computers to enable them to handle unforeseen circumstances. *(NASA)*

The Role of Artificial Intelligence

Artificial-intelligence experts have been working on this problem since the early 1960s, and most of the problems haven't been solved yet. To a certain extent, electronic computers are command override "intellectual robots" or extensions of the human mind. Today most computers are no better than the software (programs) that direct them, and continued progress in the computer field depends to a large extent on the development by human beings of better computer software, including software that will permit a certain degree of "self-programming" by a computer. Some of this work has direct impact on the studies of "artificial intelligence," which is still a field in its infancy when compared to its ultimate potentials.

However, the manner in which such software development and progress in artificial intelligence proceeds and the development times related to these are

still highly speculative, far more so than many of the activities involved with living and working in space. We may not have usable teleoperators with long time delays by the time you're out where such teleoperators or command override robots would be considered today to be optimum for space work. Thus, when it comes to teleoperators with long time delays, it probably will be cheaper to send *you* there. Or, in cases where the environment is extremely hostile—say, for example, the atmosphere of Jupiter or the surface of its moon, Io—you may be ensconced safely in a spacecraft or habitat in the immediate vicinity for the purposes of greatly reducing the time delay.

Remote Manipulators

Simple extensions to human anatomy have already been operated in space and will continue to be important in space work in the years to come.

Although extension-type tools were used by the Skylab astronauts in the 1970s to repair the Skylab, the Space Shuttle Remote Manipulator System (RMS) was the first such remote unit to be designed for moving objects around in the vicinity of the spacecraft. The RMS arm, designed and built in Canada, has a shoulder, elbow, wrist, and gripping "end effector." The RMS is visually operated by an astronaut from the flight deck of the Shuttle Orbiter, where he can watch the arm move through a window and also see the view from remote television cameras on the elbow and end effector.

The RMS is, in a way, a spacegoing crane because it is the same sort of extension to human reach and lifting power as any construction crane.

The Shuttle Orbiter "Remote Manipulator System," or RMS, arm built in Canada can be used as a "cherry picker" by an astronaut. Note foot restraints. *(NASA)*

Using tools such as this rotary power drill in weightlessness is possible if suitable foot restraints are used. *(NASA)*

Waldos

Such extensions to human reach and strength fall in the general category of devices that are often called "waldos" because these devices were first described in considerable detail by Robert A. Heinlein in a 1940 science-fiction novel, *Waldo.* In the story, Waldo Farthingwaite Jones, who suffers from myasthenia gravis, lives in the weightlessness of orbit and develops a series of power-boosted extensions to his fingers, hands, arms, etc. This was shortly before similar remotely activated arms and fingers were developed for use in handling the radioactive materials of the Manhattan District during the development of the atomic bomb.

These tools and more are, in some cases, already available for your use in space. Others will be developed, because you have yet to learn how to live and work in the weightless environment of space. Many of the activities you'll be called upon to perform out there will seem commonplace but will have some new elements introduced by the weightless condition.

Handling Solids and Liquids in Space

Handling solid objects in weightlessness is and will be no real problem. A solid object retains its shape and size. However, some solid objects are nuisances in weightlessness. Electrical wires, communications cables, and other sorts of ropes and lines are messy because they float all over, tend to get tangled up in

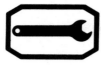

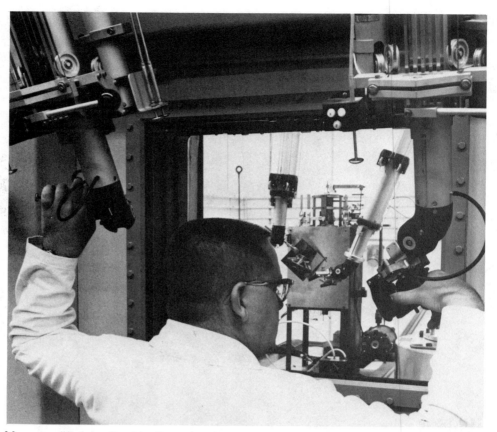

Man-amplifiers or "waldos" are used to increase human muscle strength or capabilities for heavy space work. They have been used in the nuclear industry for forty years. *(Rockwell International Corporation)*

themselves and with other objects, and get in the way. On the first Space Shuttle flight, astronauts John Young and Robert Crippen discovered that the wires connecting their communications headsets and units to electrical connections in the Orbiter were getting tangled in everything, including themselves. On subsequent flights, wireless microphones were used instead, eliminating the bothersome wires.

Handling liquids in weightlessness is something else, because a liquid, by definition, doesn't retain its shape and can easily separate itself into smaller objects. On Earth in a one-g gravity field, we're accustomed to liquid behavior of a sort that's highly unique in terms of liquid behavior in the rest of the universe. In a gravity field, a liquid will stay in an open container if the opening is "uphill" from the surface of the liquid. If the container is tipped so that the opening is "downhill" from the liquid surface, the liquid therein will stream out.

Liquids don't behave that way in weightlessness.

And you have to work to keep them under control.

Like a solid, a quantity of liquid in weightlessness will float freely wherever you put it. But, unlike a solid, it may not retain its shape. If undisturbed, a

170

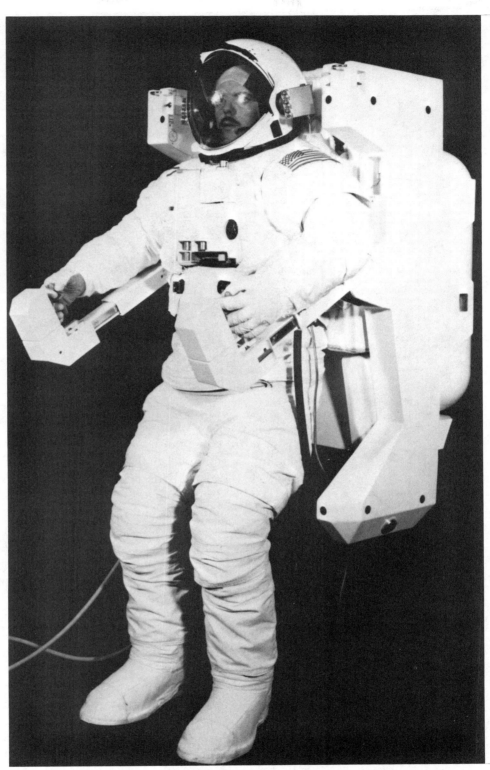

The Manned Maneuvering Unit was thoroughly tested for the first time on the tenth Space Shuttle flight in February 1984. *(NASA)*

Working in weightlessness *(NASA)*

liquid will tend to assume a spherical shape. If disturbed by any outside force such as a current of air or by internal forces caused by rotation, it becomes an oscillating semispherical glob. Disturb it to the extent that its surface tension is broken, and the glob will immediately separate into a host of smaller globs. It's very much like trying to handle liquid mercury on Earth.

Because a liquid has internal forces such as surface tension, it has a tendency to either "wet" any surface with which it's in contact or repel itself from a surface that it can't wet. Surface tension is the reason why, on Earth, if you're careful you can get a steel needle to float on the surface of a glass of water.

If a liquid can wet a surface with which it comes into contact, it will exhibit a tendency to flow out and cover the entire surface. This creates the meniscus that you can see on the surface of water in a drinking glass: the edge of the surface where it touches the glass bends upward. Only the force of gravity keeps the water from flowing up the surface of the glass to wet the entire glass, inside and out.

In weightlessness, water in a glass will crawl up the inside surface of the glass, over the lip, and all over the outside of the glass. You can't keep it in any open container that it is capable of wetting.

So liquids, in general, have to be contained and transferred in closed vessels in weightlessness.

If globs of water, coffee, or body wastes get loose in a weightless cabin, they'll either break into increasingly smaller globs until the air is filled with a mist that the life-support system will have to deal with, or the globs will touch and wet a suitable surface.

172

In general, this hasn't caused too much of a problem in space cabins because the surface tension of a liquid can be used to keep it under control.

The easiest way to do this is to mop up the liquid with an absorbent cloth, sponge, or paper tissue. Such cloths and tissues possess enormous surface areas, which liquids attempt to wet by surface tension. But once such a mop-up device is saturated with liquid, you can't just squeeze it semidry as you can on Earth, unless you have special equipment to handle the liquid you force out.

During everyday living, you won't have much trouble handling moderate amounts of liquid in weightlessness. It's only when you deal with large quantities of liquids such as may be used in several space industrial processes that big problems may arise. But, in those special cases, special procedures and equipment unique to the specific industrial operation will have been thought about beforehand and be available.

Conclusions

Performing work in the weightlessness of space may be difficult and seem to pose many new problems at first, but thus far in the history of people in space, there won't be any insoluble problems. You'll have to learn new ways of doing things, and many of the things that work on Earth, things that you take for granted, won't work the same way in space. But, since the days of Skylab and Salyut in the 1970s, there's been no question about whether or not you can perform useful work in space and no question about the ultimate value of having people in space to do work.

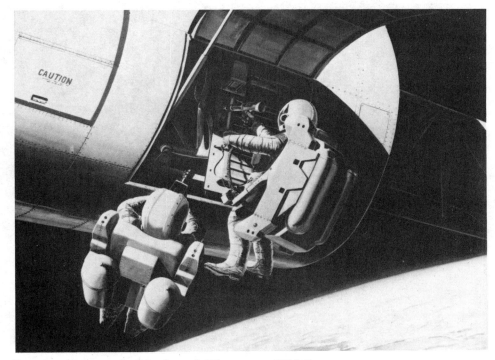

Repairing a beam builder in weightlessness *(NASA)*

11 Designing for Human Beings

The Misdesign of Human-operated Devices

People who developed the early, classic, and historic technologies that predate the Industrial Revolution paid little attention to the needs of the human beings who operated mechanical devices. At best, the design of carriages, weapons, and tools was a matter of cut-and-try—known now as "empirical development." Once some of the basic devices were developed and proven to be more or less suitable, the designs were not changed for centuries. Other devices based on these original gadgets were strongly influenced by them and therefore tended to perpetuate mistakes or misconceptions of human size, strength, or operating ability.

This principle is perhaps best illustrated by the steering of ancient sailing vessels, an operation carried out by a helmsman manipulating an oar that served as a steering rudder and, later, the hinged sternpost rudder. These steering mechanisms were located at the aft end of the hull. And so was the helmsman. When you think about it, the aft end of a vessel isn't the best place to put the man who's responsible for seeing where the ship's going and steering it. Even when blocks, tackle, and other mechanisms were introduced, it wouldn't have taken very much further development to work out steering gear that would permit the helmsman to be at the front end of the ship. But it wasn't done because the tradition of the helmsman at the rear end of the ship had been firmly established. "If it works, don't change it."

175

It wasn't until the new technology of steam power demanded a complete overhaul of ship design that the helmsman was repositioned to a bridge on or above the forecastle where he could see where the ship was going.

Imagine trying to drive an automobile from the rear seat!

This is only one example of the misdesign of devices for human operation.

Enormous difficulties are faced by farsighted advocates who try to change the basic design of a human-operated device to make it more sensible. Steam railway locomotives were operated with the engine driver located in a cab at the rear of a huge boiler that blocked forward vision. Although this was primarily because of the need for intercommunication between the engineman and the fireman, much of this requirement for communication was brought about because of the inability of the engine driver to see well. Some cab-forward steam locomotives were used on the Southern Pacific railroad from 1928 until the early 1950s, but the cab was still on the end of the boiler housing the firebox. These cab-forward locomotives were simply turned end-for-end, a technical feat made possible by the ability of the oil fuel to be pumped through a pipe to a forward firebox. It was only when the Diesel-electric locomotive came along that this major technology shift resulted in the engineman and fireman both being positioned at the front end of the locomotive. Even then, these people didn't feel comfortable without part of the locomotive in front of them, and the result of this can be seen even today in a modern Diesel-electric locomotive such as the General Electric U34 type with its cab located partway back from the front end.

But many machines were designed on the basis of tradition or ease of construction.

Why does an automobile have its engine in front? Because that's where the horse that pulled the surrey, buckboard, brougham, or coach was hitched. Designers placed the automobile engine in the front of the car to drive the rear wheels through a complex drive-shaft arrangement because it was easier to build that conglomeration than to work the bugs out of the extremely complex series of flexible joints through which power would have to be transmitted to front wheels that also had to be turnable to steer the auto. A partial solution came with the relocation of the engine to the rear of the car over the rear driving wheels, and a number of very fine automobiles were built with rear engines. Although the Chevrolet Corvair was a perfectly good car, it came afoul of an almost insignificant oversight in design, one that could easily have been corrected—were it not for an enormous amount of emotional pressure. People forget that Volkswagen built millions of safe and efficient rear-engined "Beetles" and vans over a period of more than a quarter of a century. But when technology finally arrived at the point where front-wheel drive autos were practical, almost every manufacturer began producing such cars.

The Empirical Design of Clothing

In the past, machines weren't the only things that were not properly designed for human beings. The very clothing that people wore to keep warm in cold climates, keep cool in warm locales, and provide a surrogate skin as an antichafing layer was also a matter of empirical construction.

Although some clothing, such as that developed by the Eskimos, was and is well fitted and extremely efficient, it's typical of clothing everywhere in that, to fit well, it must be individually tailored to the individual human being. This was—and still is—true for nearly all peoples everywhere. Peasants, slaves, and the impoverished lower classes wore clothing that was ill-fitting rags because tailoring was too expensive. Clothing was worn until it literally wore out. Only in tropical and semitropical climes is clothing washed; most Europeans had to learn from other people in the Orient and the tropics to wash clothes (and themselves).

You might think of people in the past as being well dressed. But the image is probably of those highly idealized portraits and other paintings of people rich enough to have their clothing made individually for them by tailors. Look carefully at the clothing of the past as preserved in such museums as the Smithsonian, and you'll see that it appears to be ill-made, ill-fitting, and extremely coarse by our modern standards.

Anthropometry and Ergonometrics

Why were all these things tolerated? Why were people satisfied with machines that were poorly designed and difficult for them to operate? Why did they put up with ill-fitting clothes?

The answer is amazingly simple: Because nobody knew how big people really were, how the sizes of their limbs and torsos varied, how far they could reach, how they sensed their surroundings, how long it took for them to see something and move a hand, the limits to their vision, and how long they could maintain top efficiency while doing some physical activity. No scientist or technician had ever measured human beings.

It was only about fifty years ago that the modern science of *anthropometry* began. Anthropometry is, literally, the science of people measurement.

An offshoot of anthropometrics is an area of technology that has only recently been given the name *ergonometrics*, or "work measurement."

People Measurement

Anthropometry has its roots in military research.

In comparison to what they are today, armies were relatively small prior to World War I. Napoleon's *Grande Armée*, probably the largest military force ever assembled up to that point in history, never numbered much more than 500,000 men at any given time. The great battles of the Napoleonic Wars in Europe rarely saw more than 150,000 men in action on each side. This was partly due to the enormous difficulty of communicating with very large military forces spread over an extensive piece of real estate. Then there was the logistics problem of supplying all these men with ammunition; they had to live off the land for their food. These armies were diverse in nature, some of them brought into battle through Napoleon's agreements with other European nations.

Somebody had to make all those uniforms. In those days, it was the work of tailors. Not all of the uniforms fit as well as contemporary artists portrayed them, and few soldiers had more than one uniform to wear. It wasn't necessary to know how big soldiers were: Officers could have their uniforms tailor-made and tailor-fitted, while the troops could be fitted out in uniforms that might not fit as well.

But, as the population increased in the nineteenth- and early twentieth centuries, mass production of clothing became necessary. Most of it was incredibly ill-fitting—just look at old family photographs for confirmation. The 1900 Sears Roebuck catalog featured mail-order clothing; but it was virtually made to order from fifteen measurements of each individual that had to be sent in with the order.

When millions of men were called to arms in Europe in 1914, it became necessary to make millions of uniforms for these soldiers. Since the task would have swamped the tailors with work, it fell to the mass-production clothing industry. Most of the uniforms were made to fit what army officers considered to be "average" men. But they had no real data on which to base the sizes of the uniforms that were ordered, nor did they really know how many of each size to order. As a result, most of the uniforms didn't fit. Or they were designed to be loose-fitting on the basis of "one size fits all." If a man was lucky, he either had a little pocket money to get his uniform tailored to fit in the nearby town or he used part of his first military pay (privates in the U.S. Army got $19 a month back then) for tailoring purposes.

The experience of the U.S. Army with poorly designed, ill-fitting, and otherwise field-ineffectual uniforms during World War I was sobering. Far too many soldiers froze to death, suffered from frostbite, or got trench foot in France because of the poor military clothing and equipment.

But for the first time the army had gathered physical data on its inducted troops. This data couldn't be used in World War I because there was little time for research in the face of war. The existence of this data led slowly in the 1920s and 1930s to the establishment of the science of anthropometry. In spite of the jokes about uniforms that didn't fit (mostly retreaded jokes from World War I), by and large the uniforms and equipment of the soldiers and sailors of the United States in World War II were far better fitting and far more effective in protecting them in the rugged outdoor environment than were those of World War I or before.

To a person from 1900, the uniforms of today's soldier, sailor, or airman would be considered outstanding examples of tailoring. As for the fantastic clothing and equipment available to the modern camper and outdoors person, Napoleon or Ulysses S. Grant would not have believed the contents of an L.L. Bean catalog.

The reason that your clothes fit so well today even though you may have bought them right off the rack in Sears, Ward's, Penney's, or any of the other big chain department stores is that now we know how big people are and how many need small, medium, large, or extra large.

178

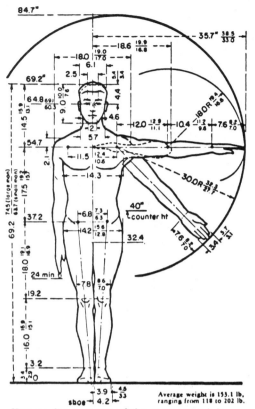

Human dimensions of the average adult male.

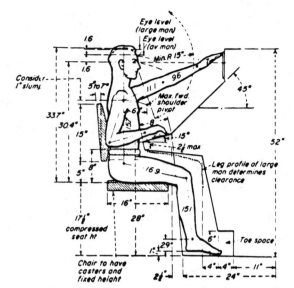

Seating is dependent on reach, vision, and work load on the operator. This drawing is based on the small man which accomodates the reach limitations of 95 per cent of all men.

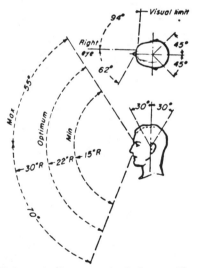

Visibility studies must include considerations of size, color, lighting, and purpose of the equipment. Where concentrated attention will be required, effective visual areas are considerably reduced.

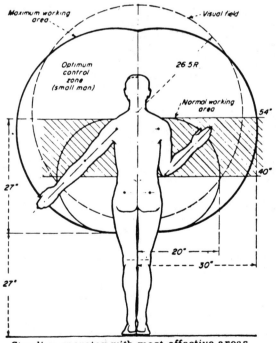

Standing operator with most effective areas for controls cross hatched.

Figure 11-1. Human dimensions (*U.S. Air Force*)

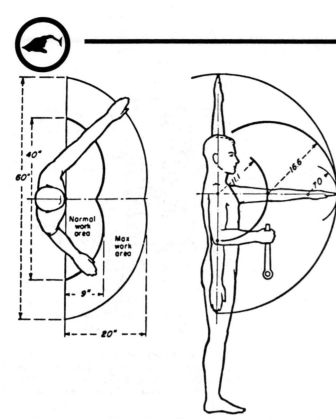

Fig. 11-1. Continued.

Reach of standing or seated operators may be extended beyond these limits by lateral motion or pivoting
at the waist.

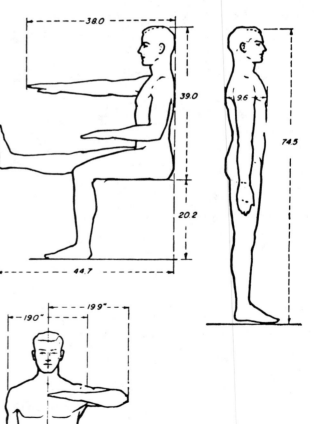

Enclosed spaces have psychological as well as antro-
pometric hazards. Space must be provided for alter-
nate body and leg positions to relieve cramped
muscles.

180

The Size of Human Beings

The drawings in Figure 11–1 illustrate how thorough this anthropometric data has become. Dimensions shown are those for a "95th-percentile man." This means that 95 percent of the current male adult American population has physical body dimensions equal to or less than those shown.

The amount of data is so large now that there are similar data bases for women as well as for children of various ages. In addition, the variation or statistical spread of the data is well known so that it becomes possible to say that so many million male adults are in each percentile of the overall adult male population.

For the first time in history, we know the size of human beings, at least in the population of the United States.

This data changes, of course. All the reasons for the changes aren't yet known. There are indications that people are getting bigger. Whether this is due to better nutrition or to genetic changes is unknown. A medieval Englishman, based on the sizes of armor, was about 5 feet, 6 inches tall. The U.S. Army soldier in World War I averaged 5 feet, 7¾ inches tall; in World War II, 5 feet, 8½ inches tall. Today's 50th percentile male army soldier is 5 feet, 9 inches tall. If the colonial houses of New England seem diminutive and quaint with their small doors and low ceilings, it's because you're bigger than your ancestors.

Using Anthropometric Data

Anthropometric data is used to design more than mere clothing, although proper clothing is an important item to you in space.

The NASA Space Shuttle space suit (which NASA calls the "extravehicular mobility unit," or EMU) made extensive use of anthropometric data. The lower torso or pantslike portion comes in a variety of different sizes and attaches by means of a beltlike metal waist ring to a hard fiberglass upper torso, which comes in five different sizes. The gloves come in fifteen different sizes, but the bubble helmet is available in only one size. Obviously, the various parts of the suit were designed with the "small/medium/large/extra-large" philosophy; while the helmet was designed to fit the 97.5th percentile—that is, it will fit 97.5 percent of the Space Shuttle flight-crew population.

Anthropometric data was also used to design and lay out the flight-crew stations. For example, complex as they may appear to be at first glance, the instrument panels in front of the commander/pilot and copilot were carefully laid out with anthropometric data. So were the various control panels at the rear of the flight deck where the payload-bay experiments and cargoes are operated.

Again, the use of anthropometric data in the design of vehicle operating positions is very new. Propeller-driven airliners had enormously complicated instrument panels because of the sheer number of factors concerned with their reciprocating engines that had to be monitored. Location of instruments was a matter of "stick it where it fits." This was also true in general aviation airplanes

181

The cockpit and instrument panels of the Space Shuttle plane look complex, but they've been designed with human factors in mind. *(NASA)*

until as recently as 1970. I once had a 1965 model Piper Cherokee whose messy instrument panel was laid out on this doctrine, and it was difficult to fly under certain conditions. I traded it for a 1969 model with an instrument panel that was completely redesigned and laid out in a logical, sensible manner reflecting not only the application of anthropometric data but information from its offshoot, ergonometrics.

Principles of Ergonometrics

Human beings who operate devices must be considered as both *sensors* and *responders* for the machine as well as the cognitive *directors*.

Treating a human being as a machine doesn't necessary mean dehumanizing the individual; in fact, by considering a person as a machinelike part of the overall system, it becomes possible to design the overall system so that its operation becomes a more humane activity.

In spite of your limited sensing range to various external stimuli—namely, your limited ability to sense various parts of the electromagnetic spectrum— you have far more sensors of far wider overall range than any other man-made device on the basis of size and weight. A device that duplicated all of your stimulus-sensation regime would be a very large and complex nonhuman robot which probably would lack another of your unique features: the ability to exercise cognitive direction without prior programming—your innate human ability to come up with answers to problems that were unforeseen and unanticipated.

As Robert A. Heinlein once pointed out in a private conversation, it would be difficult to design a robot truck driver capable of leaving the Long Beach docks and, several days later, passing through New York City's Holland Tunnel 2,600 miles away with an accuracy of inches, having in the meantime coped with millions of unanticipated decisions en route. Someday perhaps we'll be able to do it. But it may be unnecessary in the long run. In the meantime, the machines of space must be designed for people.

Human-Reaction Delay Times

In the design of machines, especially space vehicles where high relative speeds may be involved, the time delay inherent in your nervous system must be taken into account as well.

The time required for you to *perceive* the reception of a stimulus depends upon the intensity, duration, contrast, and relationship to unwanted signals than can be termed "noise." It also depends upon which sensory organ is stimulated. Threshold values below which a stimulus will not be perceived are important. A low-intensity stimulus that's below the level at which a sense organ can be activated won't be noticed. If the time duration is short, it won't be perceived: You can't perceive the dark screen that occurs between the sequential projection of frames of a motion picture, nor can you sense the 60 hertz flicker of a fluorescent lamp. If the signal has an intensity very close to that of the background, you can't discriminate it from the background; this is the principle behind military camouflage. If there's a lot of noise present, you may not be able to detect the signal itself.

As intensity, duration, contrast, and signal-to-noise ratio increase, the time required for you to sense a stimulus decreases up to the point where the particular sense organ is operating at maximum efficiency. Beyond that point, there's no decrease in perception time. At high levels, the sense organ may become overloaded. In the case of duration, the stimulus may occur over such a long period of time that you simply don't recognize it as change because the change occurs so slowly. The major things that your senses detect are *change* and *rate of change*. This holds true for other mammals as well.

Time delay within your nervous system itself is variable, depending upon the type of stimulus and how it's perceived.

Your most important sense organ for space operations, the eye, requires 20 milliseconds (0.020 seconds) to respond to a visual signal under optimum conditions of light intensity, duration, and contrast. The transmittal of this information through the nerves of your retina and optic nerve to your brain requires an additional 2 milliseconds. Once the nerve signal reaches your brain, the cortex itself requires about 13 milliseconds for excitation. Thus a total of 35 milliseconds is required simply for the *perception* of a visual signal. Under conditions less than ideal such as under stress, in the presence of high background noise, or being distracted by other stimuli or activities, this sensory

183

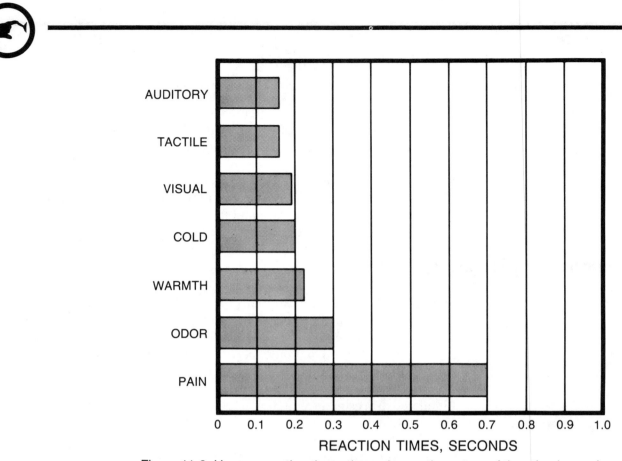

AUDITORY

TACTILE

VISUAL

COLD

WARMTH

ODOR

PAIN

0   0.1   0.2   0.3   0.4   0.5   0.6   0.7   0.8   0.9   1.0

REACTION TIMES, SECONDS

Figure 11-2. Human reaction times depend upon the nature of the stimulus and which human sensor picks it up. (Art by Sternbach)

time delay for visual stimuli can be as long as 100 milliseconds or a tenth of a second. Furthermore, the system doesn't seem able to handle a rate of data input greater than 10 bits per second, which is why you don't see the flicker of the successive frames of a motion picture being projected at 24 frames per second, the successive projections of a television raster at 30 frames per second, or the 60 hertz flicker of a light.

In this simple case used as an example, however, only a single stimulus was assumed. Because of the limiting rate of information transfer of the visual system, it can become easily overloaded by overlapping, competing, or incompatible signals. This, in turn, leads to confusion.

Responding to stimuli requires more time; how much time depends upon your physical condition, psychological set, level of alertness, motivation, and emotional state. Reaction time is also dependent upon the nature of the response—forcible movement and displacement versus precision requirements. Since most human outputs are motor responses, such as pushing buttons, turning knobs, moving switches, and activating levers or pedals, the limb that's being used to respond also affects the reaction time. Your eye-hand actions for simple tasks are about 20 percent faster than your eye-foot actions, and for a right-handed person the eye-hand reaction time is about 3 percent faster for the right hand than for the left hand.

The reaction times shown in Figure 11–2 are those for simply pushing a button in response to various types of stimuli. The usual stimuli are auditory and visual, which have total reaction times of about 160 milliseconds and 190 milliseconds respectively. Response to an odor or to pain stimuli takes much longer.

The response to a more complex task such as piloting a space vehicle during the landing phase or during a high-rate orbital closure takes considerably longer. When added to the total response time of the nonhuman system being controlled, the total time lag of the human-machine system can be as long as ten seconds or more.

Taking all time delays into account, two vehicles each having a 10-foot diameter, closing on one another at 3,000 miles per hour, will collide before either pilot can see the other, recognize the other, evaluate the situation, decide what to do, and take evasive action.

Therefore, computer systems that are much faster will be used to supplement human perception and reaction—just as they are today in high-performance military aircraft.

Human-Machine Systems

If your shortcomings as a human being are to be overcome and the optimum use of both you and the machine is desired, designing the machine around you

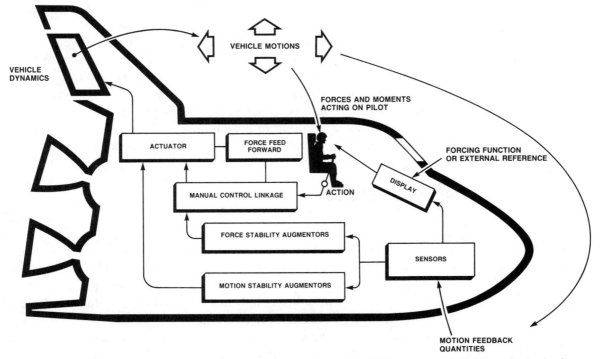

Figure 11-3. The man-machine feedback-loop-control system of a typical aerospace vehicle showing the various feedback loops. *(Art by Sternbach)*

185

is therefore an absolute necessity not only for space devices but also for other technologically advanced systems.

As we saw in the discussion about teleoperators, you must have timely information on what the machine is doing and how it's responding to your commands. In technical language, this is called "feedback," and it's the principle with which most man-machine systems operate.

A simple feedback system is shown in block diagram form in Figure 11–3. When you give a command to the machine, the machine reports that it has received your command and carries out the action commanded. It also reports back to you that the task has been initiated or completed, the extent to which the action has progressed, or other information that you must have in order to keep the machine under control. As discussed, if the time delay in the feedback loop is too long, the machine may be doing something different and unknown by the time you can get a correction command to it. Such long-time-delay systems are said to "go hyperbolic," which means that they get totally out of control and exhibit exponentially increasing departures from the desired activity.

Because of the millisecond time delays in your nervous system itself, many machine systems now include a control computer that operates in parallel with you. You can command the computer or override any computer command, but the computer does the repetitive jobs keeping the machine running properly. You then command only *changes* either to the computer or, by overriding the computer, directly to the machine.

The automotive "cruise control" and the aircraft autopilot are earthly examples of this sort of computer-assisted control. The cruise control is a computer that senses the speed of the car, compares this data against the predetermined cruise speed you programmed into the computer, then opens or closes the engine's throttle to correct the error, if any. Aircraft autopilots come in various models, ranging from those that will simply keep the wings level to those that will level the wings, hold a compass heading, track a radio navigation station, and maintain altitude. The most advanced aircraft autopilots are run by sophisticated computers and use inertial guidance systems; these devices can fly an airplane from takeoff brake release to landing rollout—but always with human operators in a command override position.

Principles of Display

Your ability to operate in a man-machine system, regardless of the degree of automation and computerization involved in the system, depends greatly on what information is displayed to you and how it's presented. This, too, is part of the subject of anthropometry and ergonometrics because it involves how you perceive things and what sort of information displays provide you with the best information under a variety of circumstances.

Spacecraft are more analogous to aircraft than to automobiles, but many of the same display principles apply to all three types of vehicles.

Automobiles in the 1950–1970 time period had very fine displays of information that was necessary to you for proper operation—speedometer, odometer, and analog or moving pointer dials presenting data on engine-coolant temperature, engine oil pressure, battery-generator current, and fuel supply. Today, on the basis of cost (it's said), most automobile instrument displays have deterioriated to digitized speed presentations plus "failure indicators," more commonly called "idiot lights," which are capable only of reporting that engine temperature, engine oil pressure, electrical system components, or the brakes have exceeded preset limits or have failed.

While it may be nice to know why your car's engine has quit on the freeway, you can always pull over to the side and stop without unduly endangering yourself or your passengers.

Aircraft are different. If something quits, it's not possible to merely pull over to the side of the road. The side of the road may be thousands of feet below and consist of rugged mountain terrain. Federal law requires that critical airframe and engine performance and condition be monitored and displayed before the pilot.

Idiot lights are the bane of an airplane pilot's life, especially in aircraft with retractable landing gear where the position of the gear, invisible from the cockpit, is reported by the opening or closing of position switches on the gear and displayed in the form of lights on the instrument panel. Gallons and gallons of cold sweat have been shed because light bulbs were burned out, switches jammed with mud or frozen with ice, or wiring connections loose. There isn't an airport in the country where a pilot hasn't made a "low pass" down the runway so that another pair of eyes on the ground could check the landing-gear position.

Types of Display

There are three basic types of information display: analog, digital, and binary.

An analog display is a circular gauge with a moving pointer such as the one shown in Figure 11–4. So is a new type of gauge called a "vertical tape," which is a circular gauge straightened out, also shown in Figure 11–4. It may also be oriented horizontally rather than vertically. A clock with moving hands is an analog time display.

A digital display is simply a presentation of a row of numbers. The mechanical odometer in an automobile is a digital representation of distance. The display on a pocket calculator is a digital display. With the enormous progress in microelectronics, America has become increasingly digitized. Digital clocks are now commonplace.

A binary display is an idiot light. It's either on or off.

Each type of display has its advantages and disadvantages.

An analog display isn't accurate; it's an indication. It may be possible to read it only to three significant numbers. However, greater accuracy than this may not be required. The important thing that an analog display can show is *relationship* of the displayed data to some maximum, minimum, or average value—the "redline" maximum or minimum, or "in the green" of average desired value. Most important, the analog display will indicate *rate of change* by the speed with which the needle or dial indicator moves. It's often important

187

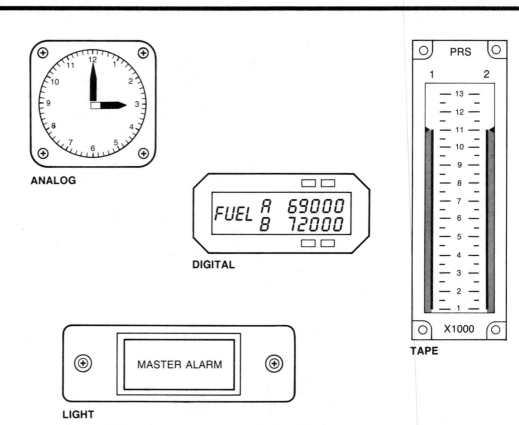

Figure 11-4. Types of displays—analog, digital, tape, "idiot lights"
(Art by Sternbach)

to know rate of change. For example, if you are driving your car up a long hill on a hot summer day, an idiot light will inform you only that the engine has gotten too hot. A temperature gauge permits you to watch the rate at which the temperature rises; if it goes up slowly, you may determine you can reach the top of the hill before the engine gets too hot, but if it goes up rapidly you know you'll have to stop partway up and let the engine cool down.

A digital display is as accurate as its sensor system and will report to as many significant figures as it has available on its presentation. It's extremely useful where high precision is required. But you must keep redline values in your head. A digital display is also relatively useless when the data changes rapidly; the display digits become a blur because they're changing so fast. An example of this is a digital stopwatch accurate to $\frac{1}{100}$ second; when it's running, you can't possibly read the rapidly changing right-hand digit.

The advantages and disadvantages of the binary display have been discussed. Any such binary on-off display requires a test mode so you can be sure that the light or display is working in the first place. It should have redundant sensors and data transmission means.

Redundancy and Triangles of Agreement

This redundancy need leads directly into the principle of "triangles of agreement." This is based upon the ancient wisdom, "If you desire to give a friend a

188

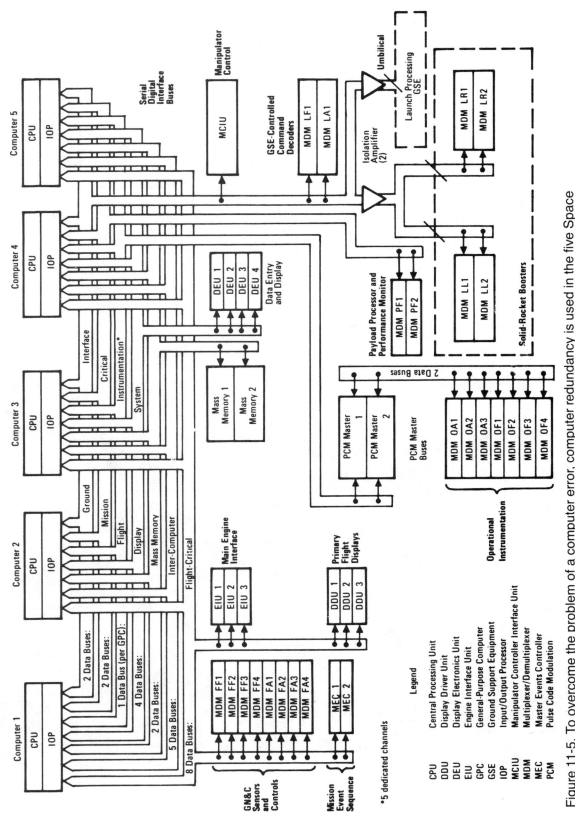

Figure 11-5. To overcome the problem of a computer error, computer redundancy is used in the five Space Shuttle computers. All five must agree before they can issue a command. (Rockwell International Corporation)

Legend

CPU	Central Processing Unit
DDU	Display Driver Unit
DEU	Display Electronics Unit
EIU	Engine Interface Unit
GPC	General-Purpose Computer
GSE	Ground Support Equipment
IOP	Input/Output Processor
MCIU	Manipulator Controller Interface Unit
MDM	Multiplexer/Demultiplexer
MEC	Master Events Controller
PCM	Pulse Code Modulation

*5 dedicated channels

189

clock, give him three so that he will know the hour." This is not the ordinary principle of triple redundancy, but something else. With one clock, you only think you know what time it is. If you have two clocks that disagree, you may not discover which one is correct until it's too late. With three clocks, you can be reasonably sure that the two that are closest in their data are probably right. Every critical control display should operate on the triangles-of-agreement principle. Every critical operation should be instrumented and displayed so that the failure of one reporting system can be covered by another related system.

Triple redundancy—"What I tell you three times is true"—requires that there be agreement between three separate systems. If one system doesn't agree with the others, something's wrong. Triple redundancy also means that it's a no-go situation unless there is agreement between three systems. In turn, for critical operations, this often means quadruple or quintuple redundancy so that you can be certain of having three in agreement.

These principles—triangles of agreement and triple redundancy—can mean that the displays for complex devices such as nuclear power plants,

Modern spacecraft and aircraft use electronic means to present data to people. *(NASA)*

Modern airplane cockpits make extensive use of human factors in the design of displays. This is an artist's concept of the new "glass cockpit" designed by Sperry Flight Systems for the Gulfstream III executive jet airplane. *(Sperry Flight Systems)*

communications switching networks, large networked mainframe computer systems, automated factories and chemical plants, aircraft, spacecraft, and space habitats can become very large, very complicated, and practically unusable by human beings because of data overload. Such wall-to-wall data may be necessary but you can't assimilate it. This has led to new techniques of data presentation and display.

Displays for Complex Devices

When you're operating even a simple machine such as an automobile, it's not necessary that all the speed, distance, fuel, and engine operating data be displayed before you constantly. It should be displayed to you in vivid format when the data falls outside a predetermined value or rate of change. Or when

191

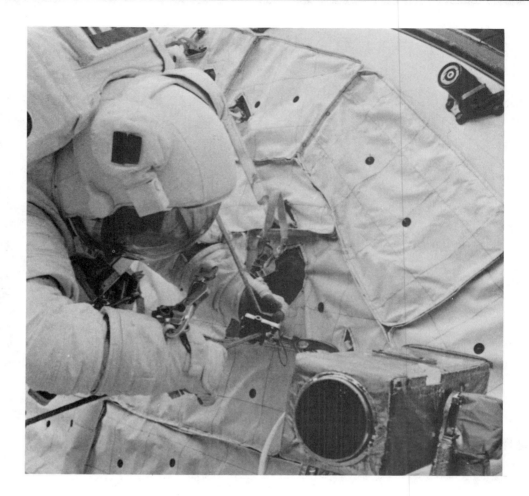

you want to know it. Or when the computer system that's tracking all this data fails in any way.

We're beginning to see this type of display now in the newer jet airliners, in military aircraft, and in the Space Shuttle itself. Information is displayed on some manner of screen in front of the operator—the commonplace cathode ray tube (CRT) will soon be replaced with other types of flat displays based on liquid crystals and other physical phenomena. The operator can select what's displayed, or the computer will automatically display any parameter that wanders out of its predetermined limits. The computer can also display recommended actions to be taken.

Control Mechanisms

Controlling or overriding the control computer has historically been accomplished by a human operator pushing or pulling a lever, positioning a control stick, pushing a button, turning a knob or shaft, or operating foot pedals. Or a combination of these movements.

Two new control mechanisms have recently appeared.

In army attack helicopters, the guns can be aimed by the pilot moving his

192

head and looking at the target. This is a complex, computer-controlled sensing of human movements other than those of the hands, arms, and feet.

Another mechanism that was first used in the NASA Gemini two-man capsules in the 1960s is commanding the machine by punching a code into the keypad of the control computer. The computer code is either an action command, activation of a new computer program, or new data for the computer to use.

Summary

Of course, all of these displays and controls are located and arranged on the basis of the foundation data of anthropometrics and ergonometrics. You can control the machines of space because we know what size you are and the most efficient ways for you to perceive and respond.

Fun in zero-g *(NASA)*

12 Recreation in Space

Although people will be in control of things in space, with computers acting as important backups, assistants, and extensions to human capabilities, there is an important difference between people and computers.

Computers can operate constantly. They don't need to stop for food, rest, or recreation. Except for an occasional period of downtime for repair, computers serve three shifts around the clock.

In spite of all the anthropometric and ergonometric techniques, people simply cannot continue to operate efficiently or effectively without rest and recreation even though they are given time to eat and take care of bodily needs.

In some of the Apollo missions and Skylab flights, the astronauts were given far too much work to do and too little time off. This caused problems, especially with one of the Skylab crews, who rearranged their work schedule to suit themselves and thereafter performed exceptionally well.

"All work and no play makes Jack . . . and ulcers." That's the modern paraphrase of the ancient paradigm.

The Need for Recreation

The military services have known for years the importance of recreation—especially in the U.S. Navy, where officers and sailors must perform at high efficiency on watch aboard seagoing vessels. The Strategic Air Command of the U.S. Air Force has also had to ensure that people can operate at top physical and mental condition during long watches in missile silos and long bomber missions. (The U.S. Army isn't faced with quite the same problems because they don't have to worry about having the battlefield sink or crash.)

All of this has a strong bearing on how you may perform under the stress of space living and working. And, although many of the results of research on human work efficiency may appear to be self-evident, the USAF Wright Air Development Center's Aero Medical Laboratory, the Holloman AFB Aero Medical Field Laboratory, and the Randolph AFB School of Aviation Medicine have done work that leads to the following conclusions regarding human performance during long and intense tasks:

1. When doing a known task, a person's efficiency starts out rather high and then drops rapidly to a plateau of performance, where it remains.

2. The more intense, bright, large, or loud a signal appears, the easier it is to detect and respond to; and there is little deterioration of performance in this regard over long periods of time.

3. The longer the watch period, the more likely that signals will be missed toward the end of the watch period.

4. Although there are marked differences in performance between individuals, the ability to stay alert over long periods of time doesn't seem to be associated with other skills or personality traits, although better watch-standing performance seems to be related to an introverted disposition.

Experimental Verification

The results of several USAF experiments confirm these conclusions.

One experiment measured the performance of a number of human subjects doing a complex task during a twenty-four-hour continuous test. Performance tended to peak during the first twelve hours, then suffered a continuous deterioration during the last half of the test.

Experiments conducted on individuals asked to perform thirty hours of continuous simple mental activity (allowing a five-minute rest each hour and twenty-minute eating periods every six hours) showed much the same pattern of performance. After about seventeen hours, performance dropped from the high plateau of the initial period of the test, reached a low point twenty-two hours into the test, recovered slightly to a brief but lower peak at twenty-six hours, and was again deteriorating at the end of the thirty-hour test.

Fatigue

Although some of the deteriorated performance exhibited in these experiments could be attributed to boredom with the repetitive task (lack of motiva-

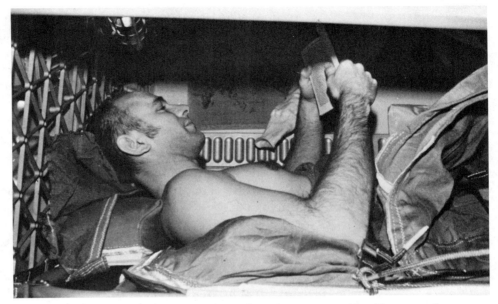

Reading in bed in weightlessness. Was Skylab astronaut Alan Bean reading Jules Verne's *From the Earth to the Moon* or another science-fiction space epic? *(NASA)*

tion), most of it was probably caused by accumulative fatigue, which is probably a far more important factor than motivation. The 1927 New York-to-Paris flight of Charles A. Lindbergh is an excellent example, because Lindbergh piloted a marginally stable airplane for more than thirty-three hours without rest, after having had little or no rest during the twenty-four hours preceding the flight. However, on the basis of experimental data obtained since that time, one could certainly speculate on how much longer Lindbergh could have flown *The Spirit of St. Louis.* He landed at Le Bourget Aérodrome outside Paris with enough fuel remaining aboard to fly for an additional ten hours. Whether the aircraft could have outlasted the pilot is a question that will never be answered.

Charles A. Lindbergh's 1927 flight from New York to Paris in the *Spirit of St. Louis* was a triumph of human endurance. *(National Air & Space Museum, Smithsonian Institution)*

Accumulative fatigue produces some strange side effects, including hallucinations. USAF School of Aviation Medicine tests had some fascinating results in this regard. In a thirty-hour extended-task experiment involving flying an aircraft simulator on the ground, the following "perceptual aberrations" were described by the individuals in their subjective verbal reports following the test:

"Occasionally the RPM indicator seemed to have a little man showing head and shoulders, in a sombrero, holding an umbrella overhead."

"At one time, the instrument panel faded out as such and became an olive-green fabric with a coarse sackcloth weave, solid in color, with no instruments."

"The needle of an instrument appeared to stand still and the face of the instrument and the whole panel seemed to revolve around the needle."

"The lower three dials seemed to be decorated as store windows—three-sided backgrounds of billowy, multicolored material with little dolls or puppets placed in each."

These are similar to hallucinations reported by other human-factors scientists investigating the effects of extreme fatigue, sleep deprivation, sensory deprivation, and drugs.

Drugs don't appear to offer any help in performing such long, fatiguing tasks. Dexedrine and caffeine produce striking initial results, greatly improving human performance for about two hours. But, over the long run, this high initial peak of performance created by stimulants rapidly deteriorates until at the end of the shift you still perform at a significantly lower level.

It's been experimentally confirmed that a good night's sleep will produce a complete recovery of your ability to perform at a high level of mental intensity. In addition, if you start work on a task soon after such a rest, you'll perform better.

The Circadian Rhythm

Another factor that enters into your ability to perform complex tasks involving decision making and judgment is the normal day-night cycle, or what is known as the *circadian rhythm* of your inner biological clock. Inside, you're a creature of habit whose activities are basically determined by the local time of day or position of the sun in the sky. There are many bodily functions that change with a circadian rhythm—body temperature, blood pressure, stomach and intestinal peristalsis, kidney and liver function, blood chemistry, and the condition of the nervous and endocrine systems, for example.

Biological rhythms of one type or another are known to exist in practically all living species on Earth, including plants. There are several natural cycles that may play a part in determining biological rhythms—the 24-hour solar rhythm, the 24-hour, 50-minute tidal rhythm, and the 23-hour, 56-minute sidereal rhythm being primary ones; with unknown roles being played by secondary rhythms such as the 14-day bimonthly, the 29.5-day synodic, and the 365.25-day annual. The extent of the interaction, or heterodyning, of all these natural cycles is yet unknown.

198

Jet Lag, or Circadian Asynchronization

The disruption of your circadian rhythm caused by moving rapidly from one terrestrial time zone to another is known as "jet lag." The two-to-three-hour jet lag of a transcontinental flight normally doesn't throw your circadian rhythm too far out of synchronization with local time. However, a six-hour transatlantic flight with a six-hour time zone change does.

Circadian asynchronization can result in a deterioration of your capability to make good decisions and affects your judgment. Its physical effects aren't as pronounced as those of fatigue although they bear some resemblance to such symptoms. Your body seems to be out of phase with what everyone else is doing. When they're ready for dinner, you're ready for bed. Or when you wake up, it's lunchtime.

An eight-to-twelve-hour rest and sleep period will permit your body to achieve circadian resynchronization, or "autophasing."

Circadian asynchronization among airline transport crews is prevented by having the crews remain on the time zone of their flight's origin during flight and during twenty-four-hour layovers for, say, a transatlantic flight.

Daily Cycles in Space

Space crews to date haven't been affected by circadian asynchronization because they, too, have continued to operate in space with the time cycle nearly that which existed at the launch site. American crews are launched on Eastern Standard or Eastern Daylight times (Zone −5 or Zone −6 times) and operate in synchronization with Mission Control at Houston (Central Standard or Central Daylight times, Zone −6 and Zone −7 respectively). Time zones are measured from Universal Time, or Greenwich Mean Time (GMT), at zero longitude, which passes through the Royal Observatory at Greenwich, England, now part of the London metroplex. Soviet crews have operated on Moscow time (Zone +2).

Habitats in Earth orbit up to and including Geosynchronous, or Clarke, Orbit 22,400 miles over the equator will probably continue to operate on the time zone of the nation controlling them. Habitats located farther from Earth could operate on any twenty-four-hour time basis they wished but will probably adopt the standard time of worldwide aviation, Greenwich Mean Time. (All flight times and weather-reporting times in domestic aviation in the United States today are already reckoned in GMT.)

Thus, because your twenty-four-hour internal clock operates on a deep biological level, you'll take it with you into space. Although some experiments on human adaptation to twenty-hour days and twenty-eight-hour days have been made, you seem to operate best on the twenty-four-hour cycle built into you by several billion years of evolution. You can't shuck that away by leaving the twenty-four-hour Earth and going into space.

199

Your inability to operate at high efficiency without a break also means that you'll take much of your current daily life-style with you.

The U.S. Navy uses a watch system known as "watch on watch" that's basically an emergency schedule of four hours on watch, four hours off watch. However, this 4–4 system can't be used for more than a week because of performance deterioration.

The NASA Skylab crews in 1973–1974 used a 16–8 system—sixteen hours of activity and eight hours of sleep, with all crew members sleeping at the same time. However, Astronaut Gerald P. Carr reported, "A guy needs some quiet time just to unwind if we're going to keep him healthy and alert up here. There are two tonics to our morale—having time to look out the windows and the attitude you guys [Mission Control] take and your cheery words."

In our highly industrialized earthly civilization—the only kind that permits space travel and living—any good foreman or manager could have confirmed those observations. In the early days of space exploration when, at best, two to three people could be placed in space for an extended period of time, it was naturally necessary to get as much work out of a space crew as possible because of the effort and expense of getting them into and keeping them in space. This changed with the advent of the NASA Space Shuttle. With crews of people living in the space station and other space habitats for months at a time, the primitive human working conditions of space exploration give way to those of space habitation.

Normal daily space living follows the 8–8–8 sequence—eight hours' work, eight hours' recreation, and eight hours' sleep. At times because of unique conditions this can go to 12–4–8, but there should never be less than four hours for "unwinding" or eight hours of sleep, except in extreme emergencies.

Sleeping in weightlessness is an experience: even the most primitive sleeping arrangement can be more comfortable than a water bed or even your grandmother's legendary feather bed. You can sleep anywhere, since there's no need to get into bed, or even to lie down.

The only real physical requirement for sleep is restraint—to keep you from floating around and banging into things, moved by the air movement in the habitat. As discussed earlier, air motion will always be present in a habitat because of the lack of normal air convection. This air movement, minute as it may seem, is capable in a six-hour period of wafting you gently toward the screened air-intake ducts along with all the other free-floating things in the habitat's compartment.

Sleep sacks are probably the best solution to sleeping in space. Built like an ordinary camping sleeping bag, a sleep sack can be attached to any surface of a habitat in any orientation. You zip yourself in. The sleep sack keeps you warm (although the habitat environment's temperature doesn't vary between "day" and "night") and also keeps you from floating around on the gentle air current of the life-support system.

Sleep sacks have been used in space since the days of Skylab in 1973.

Two cautions about sleeping in a sack in weightlessness:

Living in weightlessness. Skylab astronaut Alan Bean shaved daily with a wind-up version of an electric razor. Other male crewmen have grown beards. *(NASA)*

1. Although your eyes and brain will identify "up" and "down" within any compartment instantly, attach your sleep sack so that your head is "up" with respect to the way you think about your sleeping compartment. You may experience some momentary disorientation at any time you happen to wake up from a sound sleep; if, as you awaken, your eyes can immediately confirm that you're "right with your world," you're less likely to suffer from a short bout with motion sickness. On the other hand, one of the Skylab astronauts preferred to sleep "head down"; he said that air blowing up his nose kept him awake.

2. Don't automatically assume that you'll be able to reach out in the dark and locate something like a light switch. You probably won't. Your eye-hand coordination may adapt to weightlessness, but your kinesthetic sense may not. Like astronaut Owen Garriott, who first reported the phenomenon during a Skylab mission, when you reach out for something like a light switch in the dark, you're likely to miss by as much as 45° in angle and 8 to 12 inches in distance. It may be well to keep a dim sleeping light on just so that you can see where to reach for something *and* so that you don't suffer from a rather common nightmare: awakening to discover that you're falling in the dark.

You may find that sleeping in weightlessness is so restful that you'll require only six to seven hours of sleep each "night" instead of the usual eight that are necessary on Earth.

201

Eating in Space

The time not spent doing your job in space and sleeping must be devoted to what Gerald Carr called "unwinding."

What does "unwinding" time consist of?

First of all, you'll need more time for eating in space. As the Skylab crews discovered, the three-meal day doesn't always work out. During your initial space adaptation, you won't be able to go as long between meals. You'll become hungrier sooner, and you'll feel hunger pangs more intensely and quickly than on Earth. Be prepared for smaller meals as often as four times a day with snacking in between, even though such a schedule may be frowned upon in our "thin" civilization on Earth.

The three-meal day is historically quite recent to human experience and isn't universally followed everywhere on Earth.

Our hunting ancestors ate when they felt hungry (if they had the food available) or whenever they managed to make a kill. Your body is fully equipped to function for periods of several days without food, followed by a period of gorging yourself with all that you can eat. An example of your adaptation to this hunting existence is the existence of your gallbladder—a small, pear-shaped sac lying adjacent to and partially attached to the underside of your liver in your right upper abdomen. It connects through a system of tubular ducts to both the liver and the portion of your intestinal tract called the duodenum, just below your stomach. Its sole function is to receive and store the bile produced by your liver. Bile is essential for the digestion of fats and

People need time to relax and to socialize, even in space. *(NASA)*

fatlike substances in food. Your gallbladder stores enough bile to take care of digesting a *big* meal such as results from the kill of a deer (or maybe a woolly mammoth). The gallbladder is notorious for manufacturing gallstones, which are crystalline precipitations of cholesterol. Many people today have had their gallbladders removed because of gallstones, and they continue to eat and digest food normally. Your gallbladder therefore is a holdover from the days when your ancestors *didn't* eat regularly but had to gorge themselves when they made a fresh kill.

The habit of eating at a regular time thrice a day can be traced back to the court of the French king Louis XIV (1638–1715). Before that time in both castles and hovels, food was generally available at all times for anyone who was hungry, and people ate when they got hungry. The court of Louis XIV not only developed tableware and table manners as we know them today, but also the three-meal standard. There were so many people in the court of Louis XIV that the palace kitchens couldn't keep up with cooking for people when they wanted to eat. So there's no reason other than kitchen convenience for our tri-meal day.

Although this custom may be followed in some of the early habitats where there aren't many people and where the sociality of a meal can be as important as the caloric content, large habitats will probably operate on the twenty-four-hour cafeteria system because most space foods will, in the foreseeable future at any rate, be prepackaged ready-to-heat units in individual portions available at any time from the food locker or supply compartment.

However, the social aspects of a meal may also dictate that some of you will decide among yourselves to have regular meal hours if for no other purpose than to talk and otherwise engage in social recreation. This usually makes a meal taste better, anyway—as anyone who has had to eat alone on a regular basis will attest. Companionship and social interaction are important in space because, for some time to come, there will be few space inhabitants.

Exercise

Some recreational time must be devoted to exercise, and this is not because of any physical-fitness fad. As we've seen, your body will tend to deteriorate or atrophy in weightlessness because your muscles and heart won't have to work as hard without gravity.

Until large space habitats are built with centrifuged living quarters to provide the pseudogravity that may be necessary for prolonged living in space, mechanical exercising equipment and a definite physical-fitness period will be standard in space habitats. An hour and a half to two hours of heavy exercise daily will keep you in excellent physical condition, make it easier for you to readapt to the one-g environment of Earth on your return, and make you feel better while you're in space.

Weight lifting isn't a viable exercise medium in space, of course, because of the weightless condition of most noncentrifuged habitats. But many exercise machines are based on pushing or pulling springs; these will work in space as

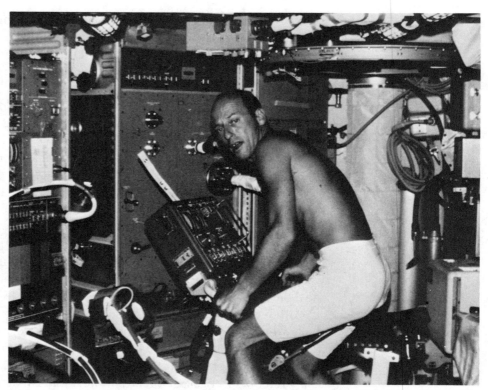

A bicycle ergometer such as the one used here by Charles "Pete" Conrad during the Skylab mission is one of the exercise devices for maintaining physical fitness in weightlessness. *(NASA)*

well as on Earth, *provided* they incorporate body restraints to keep you in place while you tug or push.

The simplest of these is the ordinary "bungee," an elastic cord or spring with handles on each end. You can use a bungee anywhere.

A bicyclelike ergonometer is useful in keeping leg muscles in good condition, although you'll have to secure yourself to the seat. Like other types of exercise, it's also good for your heart.

For those who like to run, a treadmill is probably a poor substitute, but it's about the only way that you can get a good walking or running workout in weightlessness. Unlike walking or running on Earth, you don't go anywhere . . . unless you decide to run all the way around the world—which, in a low-Earth-orbit habitat, means running in the habitat for an orbital period of about 90 to 120 minutes, depending upon the altitude of the orbit.

In some of the larger habitats with diameters of ten meters or more and open compartments that take in this whole diameter, it's possible to "run around" the entire inner circumference of the habitat. The Skylab astronauts were the first to do this. It's not running so much as it's exercise, because you must move fast enough to get a little centrifugal force to keep you against the inner circumference It's hard to keep your footing under such reduced pseudogravity, but it is good exercise and you don't have to worry too much about falling.

A bungee-type exercise device will also be used to help maintain muscle tone in zero-g. Here Skylab astronaut Alan Bean works out. *(NASA)*

Astrobatics and Other Athletics

This leads immediately into the possibility of zero-g acrobatics or, more properly, *astrobatics.* Even in centrifuged pseudogravity of 0.1 g, whole new forms of both gymnastics and dance become possible. The closest we can come to such things on Earth is on a trampoline, a three-meter diving board, or underwater ballet; each has its drawbacks in the amount of time available in the falling or buoyant state for performance purposes. Astrobatics is an extremely important form of exercise not only for the physical body itself—in terms of knowing where your limbs are and what your orientation is—but also to exercise the vestibular apparatus of your inner ears. Astrobatics may become one of the best insurances against zero-g motion sickness once the initial seizure has worn off. In their more refined and artistic forms, both astrobatic gymnastics and dance are certain to mature into new art forms as they become more sophisticated and standardized. There will also be athletic competition according to rigoristic formats; such things have happened before. In *any* area where two people can compare themselves, formalized competition has always arisen. You are part of a competitive species.

Can we look forward to the possibility of weightless "space Olympic" gymnastics? Probably yes. And, because of the outstanding communications that exist between space habitats and Earth, it's likely to be quite popular—if we can judge by the TV and motion pictures of astronauts in Skylab and the Shuttle cavorting gaily in zero-g.

Such physical activities are solo, of course, and there are other forms of

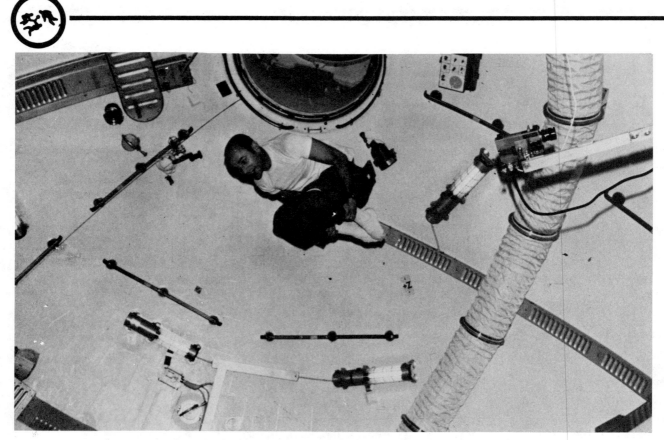

Astrobatics: gymnastics in weightlessness *(NASA)*

exercise involving more than a single person. Some of these are in the form of athletic games. The existence of reasonably large open volumes in weightlessness makes possible a whole new gamut of athletic competition.

On Earth, water polo could be played under water with a neutrally buoyant ball and with the players wearing scuba gear. But, because of the viscosity of the water, it's slow. Space polo is something else again. In a large weightless compartment with a goal located in either end and teams of players pitted against one another in the task of getting the ball to or into the opponent's guarded goal, there's speed involved in addition to the element of weightlessness. The first space-polo games are likely to have few rules other than the number of players permitted in the compartment at any one time. Basically, the original rules will be something like, "No kicking or gouging, biting or clipping. No holding for more than two seconds. Come out flailing, and may the best team win." Football helmets may be required because in such a weightless melee it's possible to get kicked in the face or head. Throwing the ball itself results in the thrower exhibiting a reaction force unless he's anchored against a wall. Personal motion is the result of pushing against a wall *or* against another player, which act also causes the pushed player to go sailing off in the opposite direction. There are injury possibilities involved with getting kicked, being involved in interpersonal impacts, and colliding with walls. We don't know what the extent of such injuries may be yet, and we don't really know what sort of personal protection in the form of equipment will be necessary. A helmet and face guard are certainly among the first considerations, however. Knee and elbow padding as well as wrist and ankle wrapping may be necessary. But there will be injuries, and reasonable attempts should be made to keep these at a minimum.

In time, such weightless athletic games as space polo may compete with professional football, baseball, basketball, hockey, wrestling, and boxing on television because it could be an interesting synthesis of all of these.

Even handball, squash rackets, paddle tennis, and tennis itself become totally new and fascinating games when played in the enclosed volume of a weightless habitat.

All of these are going to be strong, active, and even violent competitive sports because human beings with their hunter ancestry are competitive animals. These strong competitive athletic sports in space will evolve as outlets for the physical and mental violence that's part of each of us.

Initially, there may be some international space competitions in this sporting area. But perhaps not. Perhaps the biggest competitions may well be and should be between those who live in different habitats. And perhaps this sort of inter-habitat sporting competition started early in the human habitation of space may do a great deal to defuse what many people believe is the greatest human hazard in space: international competition.

Once you have competed personally against someone from somewhere else, that person seems neither so fearsome nor so fearful, whether you really like him or not.

Social Interaction in Space

All social interaction in space isn't competitive, however. There are competitive and noncompetitive social activities on Earth, and it's no different in space.

You'll take your social graces into space with you because such things as manners and protocol are important. They're integral parts of your social mechanism.

Any and all human social organizations have what the term itself implies: organization. The basic nature of a social institution is its internal structure and organization. As we'll see in chapter 15, space living offers us some new opportunities in this direction. But you'll always have to know who's in charge of what and what responses are expected of everyone in most social interactions. This is simply the matter of manners. Since deep inside each of us is a territorial hunting animal, manners serve as a lubricant in interpersonal contacts. Our covetousness and other aggressive traits derive from the evolution of our species in a world of apparent scarcity where there seemed always to be not quite enough to go around for everyone. We're learning that this isn't true. But, in the meantime, we're still saddled with these ancient, ancestral traits developed for survival in a preindustrial world.

There will, of course, be a certain camaraderie in space because there's a common physical enemy always present behind that pressure bulkhead. But

207

inside the pressure hull, you will experience human contacts and the inevitable human problems, especially during recreational times.

To a small extent—probably because of a combination of the unique experience and high motivation—some of this was present in the early-day space crews. The Soviets reported some personal problems that developed among their two-man *Salyut* crews during very long space missions. They also existed in American crews but were generally deemphasized by both the crew members and NASA.

There will be space parties—those social gatherings of a recreational nature where people do as they've done on Earth for centuries: gather together to talk, eat, and play. Americans have celebrated the Thanksgiving holiday in orbit. The Soviets have celebrated crew-member birthdays in space. Such gatherings for celebrations are an integral part of our hunting heritage and are absolutely essential for getting along together in space just as they are on Earth.

Men and Women in Space

Save for special missions by the Soviets that may have been intended more for publicity purposes than physiological study, the early space farers were male. This is probably because, due to the experimental nature of the spacecraft, most early space crews had to be highly skilled aircraft test pilots. In spite of the fact that there have been some outstanding female pilots, such as Jacqueline Cochran, Sheila Scott, and Jacqueline Cobb, there were far more male test pilots available for selection as early astronauts. However, we've witnessed a healthy general change in social attitudes and moral restraints as well as improvements in spacecrafts. The Space Shuttle and the Soviet Soyuz spacecraft opened the window of opportunity for women to travel into space.

There have been women in space ever since, and there will always be women in space from now on. This is as it should be. Space is, after all, more than a place where pilots can test their mettle by showing the right stuff, and it's more than a place where scientists go to pursue their favorite hobby: scientific investigation. It's a *human* activity, which the Soviets have acknowledged even from the prepioneering days of 1903 when the Soviet pioneer Konstantin Eduardovich Tsiolkovski wrote, "Earth is the cradle of reason, but one does not remain in the cradle forever."

The presence of women in space naturally leads to consideration by authors, space buffs, and even the aware general public of the possibilities of our favorite recreation carried out in the space environment. It has to be our favorite recreation because we're all basically obsessed with sex, Sigmund Freud notwithstanding. It probably stems again from our ancestral hunting mammalian heritage where sexual activity was directly linked to reproduction, the strongest of all survival drives, even stronger than self-preservation, because it amounts to the only way we could achieve any semblance of immortality in a mortal world.

The concept of romantic love grew slowly from this basic instinct, and the great upheaval of transition from a have-not hunting and agricultural civilization to a superindustrial world of abundance led in turn to an evolution of sexual activity to what amounts to either the greatest recreation of all or, in some cultures, an art form of its own.

Of course we'll take sexual activities into space with us! We're doomed if we don't. (And, according to some prudes, damned if we do.) If, in the long haul, we wish to colonize space, that activity requires men and women reproducing themselves as well as enjoying themselves. The first actual space colony won't be constructed; it will be created in any space habitat when the first child is born there.

As of this writing, human sexual activity in space hasn't *openly* happened yet. But it will.

Certainly it can be done in weightlessness. Gravity isn't necessary, and this contention could be proved (and perhaps has been) by experiments conducted in neutral-buoyancy water tanks. We really don't need to conduct those tests, however, even though funding to carry them out might be obtainable from a variety of private sources. Our remote ancestors evolved in the neutral buoyancy of the Earth's oceans, and more recent ancestors who left that ocean to live on land had to learn how to perform sexual activities in gravity. Those species who did so managed to survive.

Therefore, in space it's going to be even more of a matter of doing what comes naturally because, in essence, we're returning to the neutral-buoyancy conditions of the Earth's oceans!

Each of us must rely on personal experience to a large extent in considering this because each of us has had a definitely personal experience in this area that's probably unlike that of anyone else. Sexual activity is unique in human experience because, in spite of books and manuals and even personal coaching, until you've done it, you don't know anything about it.

There doesn't seem to be any reason that sexual activities of most sorts can't be adequately and satisfactorily carried out in weightlessness. Thanks to our ancestral heritage of tree living, we're all equipped with arms and legs that can grasp. Therefore, any action and reaction forces generated can, with some experience, be perfectly controlled to maintain intimate contact.

Your initial experience in weightlessness, especially if you've recently arrived from one-g Earth, should probably take place in a sleep sack where there are some physical restraints to prevent uncontrolled forays around the compartment.

Judging from the plethora of ancient temple carvings as well as modern printed activity manuals, there isn't a single one of the legendary Thousand-and-one Ways of the Ancients that won't work in the weightlessness of space.

It's really not necessary to be explicit. Nor is there any reason to be apprehensive. You'll do fine. Fish do, and they don't have arms, hands, and legs.

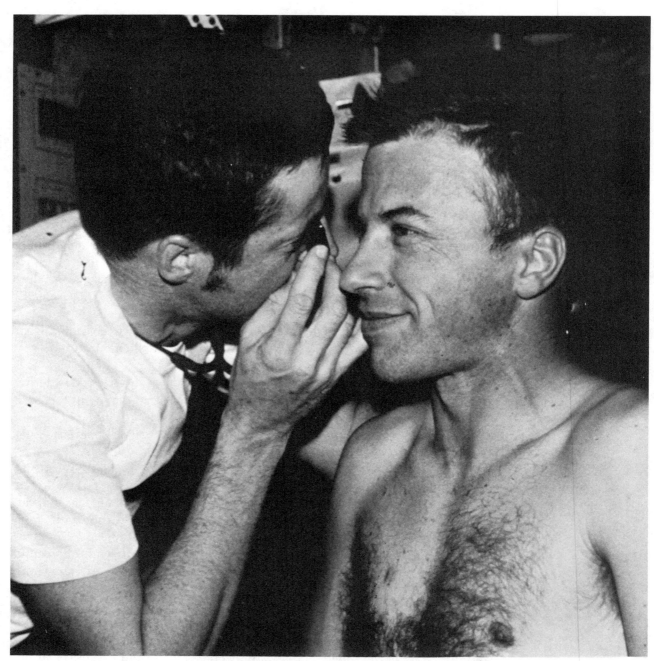

Dr. Joe Kerwin was the first medical doctor to practice in space in the Skylab. *(NASA)*

13 Health and Medicine in Space

Preflight Medical Requirements

You won't be permitted to go into space in the first place unless you're in good health and suffer no serious congenital ailments. However, as the number of people living in space increases and as more background information and experience is acquired concerning true space medicine, many of the stringent medical requirements initially imposed will be relaxed. This has been the general trend in space travel.

The first seven Mercury astronauts had to be in top physical condition. The medical examinations given to these super-healthy test pilots at the Lovelace Clinic in Albuquerque, New Mexico, were perhaps the most extensive ever devised. As concerns about the physical effects of the space environment on astronauts slowly disappeared in the light of growing experience, the medical qualifications and standards were relaxed. By the time the NASA Space Shuttle became operational in the early 1980s, the required medical examination was basically no more difficult or extensive than an FAA Third Class Airman's medical examination, which is passed every two years by more than three quarters of a million private pilots in the United States.

In the past, astronauts have lost their chance to participate in a space mission because of temporary minor illnesses or possible exposure to contagious communicable diseases. However, even with these strict precautions enforced by a preflight physical exam, the three-man crew of the week-long *Apollo 7* mission all came down with head colds *after* getting into orbit, leading their waggish commander, Wally Schirra, to call *Apollo 7* the "Seven Day Cold Capsule." Thomas Mattingly was bumped from the position of Command Module Pilot on *Apollo 13* because he'd been accidentally exposed to measles a few days before the launch . . . and therefore didn't participate in the near tragedy of *Apollo 13*'s aborted lunar landing mission.

If you've got a chronic or congenital health problem, you may have to wait to go into space until the health standards are relaxed somewhat. However, the list of potential medical disqualifications appears to be short even now.

Basic Medical Requirements

The medical requirements for space travel will probably follow those of the current Federal Aviation Administration 1983 standards of physical eligibility for a Third Class Airman's Medical Certificate for student and private airplane pilots:

Eyes. Distant visual acuity of 20/50 or better in each eye separately without correction or, if corrective lenses are worn, 20/30 or better. No serious pathology of the eye.

Ears. Ability to hear the whispered voice at 3 feet, and no acute or chronic disease of the inner ear. No disturbance in equilibrium.

Mental and neurologic. No established medical history or clinical diagnosis of any of the following: (1) a personality disorder that is severe enough to manifest itself by overt acts; (2) a psychosis; (3) alcoholism, meaning a condition in which alcohol has become a prerequisite to a person's normal functioning; (4) drug dependence, as evidenced by habitual use or a clear sense of need for the drug; (5) epilepsy; (6) a disturbance of consciousness without satisfactory medical explanation of the cause; (7) a convulsive disorder, disturbance of consciousness, or neurologic condition that might endanger the person or others.

Cardiovascular. No established medical history or clinical diagnosis of myocardial infarction, angina pectoris, or other evidence of coronary heart disease that may reasonably be expected to lead to myocardial infarction.

General medical condition. No established medical history or clinical diagnosis of diabetes mellitus that requires insulin or any other hypoglycemic drug for control. No other organic, functional, or structural disease, defect, or limitation that might affect the person's ability to perform safely the activities of living in space.

Naturally, there will be individual cases in which various sorts of waivers will be granted; any medical standards that are this generalized can't be as closely applied as those of the First Class Airman's Medical Certificate required for airline transport pilots. Individuals can and do deviate from these general norms and can still be considered healthy and capable.

For example, many private airplane pilots have obtained waivers for the cardiovascular requirements because, although they've suffered heart attacks, they've undergone bypass surgery or have undertaken rigorous rehabilitation programs to recondition their hearts. There are handicapped pilots, such as those belonging to the Wheel Chair Pilots Association, who have obtained the necessary waivers to be able to fly even though they may be, for example, partially paralyzed.

As the years go by, more and more waivers and deviations from the medical standards and requirements will be given. Space conditions themselves may turn out to be far more benign for certain handicapped people.

Communicable Diseases in Space

If you're suffering from an acute or chronic communicable disease, however, it's going to be a different story. Space habitats are the cleanest and healthiest of all human living situations. They must be maintained that way for a number of reasons.

In the first place, it costs money to maintain you in space. You're there to do a job that a machine can't because you can do it better and cheaper. If you become ill and have to take sick leave, the cost of maintaining your life support continues during your period of illness or incapacitation. Contingencies such as temporary illness will be taken into consideration in the design and manning of the habitat, so that a few people on sick call won't disrupt daily operations. It becomes a serious matter, however, when a communicable disease triggers an epidemic in a small and tightly controlled space habitat.

A similar operational philosophy exists aboard ships on the Earth's oceans or in the frontier living conditions of the North Slope of Alaska or the scientific stations in Antarctica.

If an epidemic breaks out in Prudhoe Bay or Byrd Station, help is only hours away by air if the weather cooperates. The sick can be flown out, or immunizing agents and antibiotics can be flown in. The analogy holds true for most of the space habitats in low Earth orbit, but it may not hold for those in geosynchronous orbit and certainly not for those in lunar orbit or on the Moon. The greater the distance from Earth or from a major medical facility, the longer it takes to bring in help or to get the sick and injured to better facilities, if such facilities exist and if the sick and injured can be moved.

This is why in the early years of space habitation extreme measures will be taken to prevent communicable diseases from getting into space habitats. If you've got a head cold, you're going to have to wait until you get well before you can go to space.

Quarantine

In order to gain entry into some habitats, especially those heavily involved in the space manufacture of biological materials where exceptionally stringent standards have been established, you may be quarantined for a period of up to several weeks until it can be absolutely ascertained that no undesirable micro-organisms are present on or in your body, until your intestinal flora have been flushed out and replaced with those more benign to the biological operations taking place in the habitat, or until you can be certified to have physiologically and pathologically adapted satisfactorily.

Other habitats, especially those having a total or partial military function, may require a quarantine in order to prevent the possibility of the introduction of chemical- or biological-warfare weapons. Space habitats are exceedingly vulnerable to a wide variety of these.

Space-made Pharmaceuticals

However, because of the biological materials that are made in the weight-lessness of space, it's very likely that immunizations and specifics for nearly every known disease, including the common cold, will be available very early in space habitats. These space drugs—you may be involved in the manufacture of some of them—will ensure that space habitats remain the healthiest of all human environments and will do much to stem the tide of disease and sickness on Earth where, because of the larger biosphere, it's more difficult to do the same thing.

The problem of communicable-disease epidemics in space will probably be solved by biological materials made in space.

Personal Hygiene in the Habitat

However, personal hygiene will continue to be important in space habitats regardless of whether or not biological specifics become available for most communicable diseases. The military services long ago discovered that per-sonal hygiene is extremely important when groups of people live and work in close proximity to one another. Because of the requirements for good personal hygiene, space habitats will make most Earth cities and living places seem incredibly dirty in comparison.

Personal hygiene means keeping yourself clean and keeping your immedi-ate living and working environment not necessarily sterile clean, but hygienically clean.

Any medical pathologist will tell you that your body is a breeding and feeding ground for an extremely wide variety of microorganisms, some of which are pathological and others of which are benign. You live in a symbiotic or cooperative relationship with thousands of other terrestrial organisms. Sometimes organisms that are symbiotic, harmless, or helpful to you can be hazardous to others, because we're all individuals with slightly different genes and body chemistries. Earthbound hospitals are extremely aware of this fact, especially in those parts of their operations where cleanliness must be main-tained. This is less a matter of getting you and your environment clean in the first place as it is keeping it clean. The same holds true of a space habitat.

214

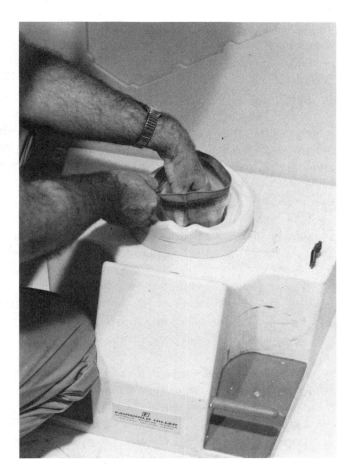

The Skylab waste-management collection device, otherwise known as a space toilet, utilized air movement to pull wastes into a removable plastic container that was later taken out and sealed. Other habitats will use devices similar to the Space Shuttle Orbiter's toilet. *(NASA)*

Microorganisms and viruses enter an earthbound hospital in the air and in supplies and on people entering the facility from the outside world. In space, the only source of microorganisms will be incoming supplies and people. There are ways to reduce the inflow of organisms in and on supplies, but there is absolutely no way of eliminating the influx of those organisms, pathologic and benign, that come on and in people because each of us is alive and crawling with such organisms and we can't really do anything about it.

In addition to the combination of natural human microorganisms are the scents and odors that you exude as part of perspiration, which keeps you cool, and as natural scents such as pheromones. All of this means that, in spite of the life-support system, a space habitat will grow to smell like the inside of a locker room after a tough game if you and others aboard don't maintain good personal hygiene individually and collectively.

Personal-Hygiene Kits and Activities

The design and function of some space habitats or the nature of some spacecraft and space missions will be such that you may be provided with a standardized personal-hygiene kit such as the ones pioneered in the NASA Space Shuttle

215

flights. A PH kit for men might contain, for example, a week's supply of personalized material such as a razor, shaving cream, and styptic pencil, skin cream or emollient, stick deodorant, nail clippers and file, comb and brush, dental floss, toothbrush, toothpaste, antichap lip balm, and soap.

The operational requirements of long-duration space habitats with advanced life-support systems may allow you to make up your own personal-hygiene kit from a list of commercial products that have been selected primarily for their compatability with the life-support system. You may also be allowed a limited selection of other small personal items. Replacement, refurbishment, or refill of your personal items will be accomplished by ordering them sent up from Earth in a cargo payload.

Stay personally clean. Personal-hygiene measures in space are no more stringent than those necessary when living in any close quarters on Earth, with the exception that you can't possibly get away from the habitat on a moment's notice.

Wash your face and hands regularly. Take a shower as often as permitted and possible. If a shower isn't possible, take a sponge bath. Don't depend upon antiperspirants to knock down personal odor; they won't work very long and actually restrict the act of perspiration you may need to maintain a proper body temperature.

Wear clean clothes and when clothing gets dirty, send it for cleaning. Doing laundry in space may be out of the question in the early days of space habitation. There may be no facilities for it. You may have to do what the forty-niners in San Francisco did: send your dirty laundry somewhere else for washing. In 1849 dirty laundry was actually shipped by sailing vessel all the way from San Francisco to Hawaii to be washed and then returned, because San

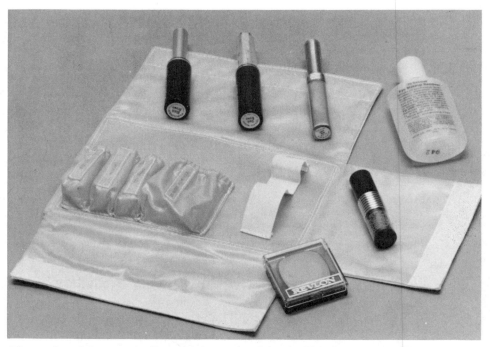

The optional personal-effects kit issued to female Shuttle crew members *(NASA)*

Francisco didn't have the available water supply or the people available to do the laundry in the first place. You'll probably have to ship your dirty laundry back to Earth for washing.

Waste-Management Hygiene

You may suffer initially or even occasionally from motion sickness brought on by the new environment of weightlessness. It's important to anticipate that you may vomit. Carry a sick sack with you and *use it*. Otherwise, you'll probably have to clean up the mess yourself. In addition to being an unpleasant task, it doesn't help others around you who may exhibit sympathetic physical reactions brought about by the appearance and odor. Remember that liquids go all over the place in weightlessness and are difficult to get under control once they've escaped into the habitat.

It is equally important that you maintain strict personal hygiene when using the various types of zero-g toilet facilities. Most of these units operate on the same principles as those used in the Skylab and Space Shuttle. The force of gravity is replaced by the flow of air to draw urine and feces into collector enclosures. The more advanced life-support systems reprocess the urine by extracting the water from it and storing the residue as solid waste. Fecal collectors utilize a similar principle to separate water from the solid matter to which antibacterial agents are added to inhibit the growth of bacteria and eliminate the possible release of odors. The solid wastes from urine and feces are vacuum-freeze-dried in containers that are replaced from time to time with new ones, and the used ones returned to Earth for disposal there. In the more advanced closed-cycle life-support systems that must be used in the more remote habitats, this solid waste is recycled by the system into nutrients for the habitat farms that not only help recycle carbon dioxide into breathing oxygen but also provide a limited amount of fresh greens for food.

Living in some remote habitats or other spacecraft without dedicated toilet facilities will require that you use the modern version of the old Gemini Bag.

You may have trouble adapting to some of these techniques of personal waste management in space because of early cultural and toilet training on Earth. In space, you'll have to overcome some of the distaste or other psychological blocks you may have or you may find living in space extremely difficult.

The most useful item for controlling both liquid and solid matter involved with personal hygiene is the disposable moist towelette. Some of these are available in sealed foil packages that you can carry around with you for emergency use. Don't carry a used towelette around with you for long periods of time. Be sure to dispose of them properly and promptly in the nearest toilet facility.

Waste Management in a Pressure Suit

Handling waste management in a pressure suit is another matter. Pressure suits have built-in urine collectors, and some of them also have built-in versions of

217

the Gemini Bag mentioned earlier. When suiting up, it's important to ensure that either the male or female urine collector is securely attached to your body so that there's no possibility of leakage. When desuiting, be careful to seal the sphincters and close the valves to prevent the escape of any wastes. In some facilities, space-suit technicians will take care of removing the collection systems and disposing of any material in them. However, in most habitats this is a task you must do yourself.

If your job requires that you spend long periods of time in a pressure suit every day, you may develop a personal diet and hygienic regimen that permit you to spend as many as six hours in a pressure suit without having to relieve yourself. If you're able to do this, it's preferable to having to use the suit waste-collection systems. However, you should *always* attach them so that they're available for you to use in case of emergency. It's impossible in a pressure suit in vacuum to attach a waste-collection system that you neglected to hook up during suit-up under pressure.

Space Medicine

Although "space medicine" was recognized as a definitive branch of the medical profession long before the first men flew in Earth orbit, it was essentially an offshoot of aviation medicine and, as such, was primarily concerned with research into human factors and physiological responses to the aerospace environment. True space medicine—the prevention and treatment of injuries and illness in space—didn't start until people actually began to live in space for weeks at a time. Now space medicine is an active area of the medical profession. Physicians have practiced in space and will continue to do so in the future, as more and more people live in space to carry out an increasing variety of scientific, technical, construction, and support tasks.

There are eight basic areas of space medicine:

1. Trauma medicine concerned with job-related or accidental injuries of a physical nature such as cuts, burns, bruises, abrasions, broken limbs, and even partially or totally severed limbs.

2. Pathology, especially that concerned with personal hygiene, "public health" in a spacecraft or space habitat, and preventive and anti-epidemic measures dealing with infectious agents such as bacteria, fungi, and viruses.

3. Treatment or amelioration of suffering related to congenital afflictions such as appendicitis, tonsillitis, cholecystitis, earaches, toothaches, and possible allergic reactions to some types of materials used in spacecraft and space habitats.

4. Treatment of stress-related illnesses manifesting themselves in hypertension, cardiovascular problems, and even psychosomatic conditions.

5. Handing of psychological problems caused by the unnatural environment of a space habitat and phobias that may not manifest themselves

until a person is in the space environment for several hours or days. These might include claustrophobia or acrophobia; they've surfaced even among the early, highly trained, and highly motivated astronauts.

6. Biochemical problems brought on by dietary deficiencies, glandular imbalances, blood-chemistry difficulties, the lack of trace elements in the food, etc.

7. Medical problems caused by social interaction. There will be conflicts, regardless of how well people appear to get along with one another, and regardless of rules, regulations, and security measures, someone is certain to smuggle alcohol or drugs aboard to liven up what they consider boredom.

8. The basic medical problems unique to the space environment itself. These include hypercalcemia due to calcium resorption; the Kittinger Syndrome or "vac-bite," caused by accidental exposure of part of the body to vacuum because of a suit failure; pneumo-poisoning caused by an excess of such gases as oxygen, nitrogen, carbon dioxide, and carbon monoxide caused by life-support system failure; and even "traumatic abaria," or the partial or momentary total loss of ambient pressure. Included among these space-related medical problems is the physiological and psychological effects of ionizing radiation, especially in those habitats located beyond the Earth's Van Allen radiation belts, where the effects of ionizing radiation from solar flares become a serious health problem.

Contingency Medicine or "First Aid"

The portion of space medicine that you're most likely to come into contact with on a personal basis has been termed (in the usual aerospace parlance of equivocation and ellipsis) "contingency medical activity." It's easier to call it by its old, well-known name: first aid.

Because physicians and medical facilities may not be available in many spacecraft and in some of the smaller space habitats, and because hours or even days may pass before professional medical help can be obtained, you'll be thoroughly trained in first-aid techniques to use for minor illnesses and accidents and to stabilize the physical condition of someone who's been severely injured.

In common with commercial airliners, all spacecraft carry first-aid kits containing provisions for emergency medical treatment and relief of suffering for minor illnesses and injuries. Some complex kits in deep-space craft also have provisions for stabilizing severely ill or injured people until professional medical assistance becomes available. Flight crews are trained in the proper use of the contents of the kits, but you may be called upon to administer first aid if the flight crew is busy with other vital duties.

Other types of first-aid kits are located at critical places in space habitats, such as at or near air locks or in the vicinity of equipment whose malfunction

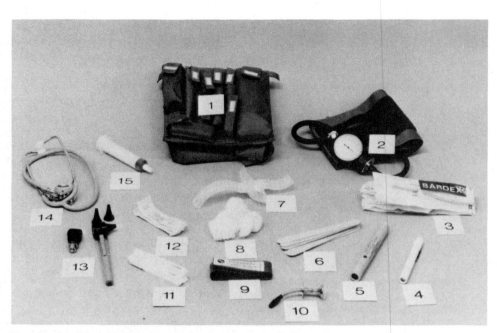

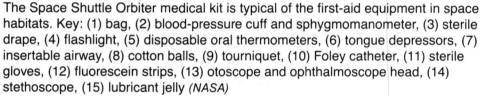

The Space Shuttle Orbiter medical kit is typical of the first-aid equipment in space habitats. Key: (1) bag, (2) blood-pressure cuff and sphygmomanometer, (3) sterile drape, (4) flashlight, (5) disposable oral thermometers, (6) tongue depressors, (7) insertable airway, (8) cotton balls, (9) tourniquet, (10) Foley catheter, (11) sterile gloves, (12) fluorescein strips, (13) otoscope and ophthalmoscope head, (14) stethoscope, (15) lubricant jelly *(NASA)*

could cause a personal injury. All space habitats have either an area specifically set aside as an infirmary or hospital, depending upon the size of the habitat, or an area that can be quickly converted into a medical treatment or infirmary compartment. Part of your initial orientation in any space habitat will be a briefing on the location of first-aid kits and facilities, where the "sick bay" is, and the communications procedures to be used in calling for help from medical personnel in the habitat or elsewhere.

All first-aid kits contain more or less the same materials, although more complex units may be used in spacecraft or remote habitats without dedicated medical capabilities. There may be as many as three plainly marked packages in a first-aid kit:

1. A medications packet containing oral medications, injectable medications, and noninjectable medications. Some of these may not be administered without approval of a paramedic and, in some cases, a physician. However, in situations where a paramedic or doctor isn't immediately available but can be contacted by radio, oral instructions for use of such medically restricted items may be given and are considered to constitute approval or permission.

2. A bandage and equipment packet containing Band-Aids, Steri-Strips, tape, gauze, wipes, swabs, injectors, syringes, intravenous injection

equipment, tweezers and forceps, scalpels, scissors, hemostats, sutures and needles, sponges, and splints. The contents of any given kit will depend again on where it's located and the availability of medical personnel and facilities. Some kits may contain equipment that cannot be used except by trained professionals such as physicians, nurses, or paramedics or that may be used by nonprofessionals provided communication is established with qualified medical personnel and the equipment used with their approval and/or oral supervision.

3. A diagnostic/therapeutic packet containing a stethoscope, blood-pressure cuff and sphygmomanometer, airway, thermometers, penlight, catheters, otoscope specula, ophthalmoscope, binocular loupe, sterile drape, sterile gloves, and defibrillator. Here again, the type and nature of the equipment in this packet depends on where the first-aid kit is located. The use of some of the equipment must be carried out by or under the supervision and with the approval of qualified medical personnel.

All first-aid kits must have a complete list of their contents plus a quick-reference handbook giving explicit instructions for the administration and use of each item in the kit. Those medications and items whose use requires professional presence or remote oral approval and supervision are sealed in packs that are plainly and clearly marked regarding these restrictions.

The complex first-aid kits containing certain drugs and equipment are the responsibility of a named person in the craft or habitat who regularly checks the kit inventory. If the seals are broken, the person responsible for the kit must account for the drugs and equipment that have been taken from the kit.

First-Aid Techniques

You'll encounter accidents as well as sickness. Not only is there no reasonable way to prevent all pathogenic agents from entering a habitat, but it's also impossible to perform any sort of work anywhere without the possibility of an accident. It's impossible to protect the world from idiots or to protect idiots from themselves. The only practical approach that engineers can take within the bounds of reasonable costs, efficiencies, and reliabilities is to design equipment, tools, and other devices so that it's difficult to operate them incorrectly or to get injured by them. Managers and personnel evaluators must then hire people who possess the intelligence, discipline, and expertise *not* to get hurt in the first place. Space is no place for the accident-prone person.

However, in spite of the finest engineering and all of the protection techniques required by regulations as well as prudent management, people get hurt and will continue to get hurt on the job. Space is no exception, because in this regard it's another work environment. Every engineering job and industrial operation on Earth proceeds with an estimate of anticipated injuries and even deaths, and the same holds true in space.

Be prepared for accidents, and be prepared for *unusual* accidents because, for all the research and development that has gone into space living, space is

221

still a hostile environment and one of the most deadly ones yet broached by human beings.

First-aid techniques in weightlessness are unique and cannot be duplicated or taught on Earth. Therefore, much of your first-aid training will be conducted during your orientation in the habitat itself.

For example, intravenous injections cannot be administered by the drip method in weightlessness because there's no gravity to transport the fluids; special positive-displacement intravenous injectors resembling very large syringes are used. Some of these are manually operated and require very careful operation to ensure that the intravenous injection is properly metered into the patient. Others are battery-powered and can be adjusted to provide a calibrated injection rate. Some of the training in this technique will take place on Earth, with final training being given in weightlessness.

It's extremely important that you learn how to properly insert an airway and that you do so as quickly as possible when administering first aid to an unconscious person in weightlessness. To an even greater extent than on Earth in a one-g gravity field, it's possible and even highly probable that an unconscious person will either swallow his tongue or inhale regurgitated stomach fluids. Insertion of an airway will prevent such things from occurring. The rapid and timely insertion of an airway is therefore extremely critical *whether or not it appears that an airway is needed at the time.*

Space Medical Facilities

Large habitats and habitats located in geosynchronous orbit or beyond will usually possess well-staffed and completely equipped hospitals, although some of them may appear to be so small that they deserve only the title of "infirmary." However, even the smallest and most austere space hospitals possess the basic equipment for diagnostic purposes as well as facilities for carrying out the most complex and advanced surgical techniques.

Hospitals in habitats have excellent communications links with other facilities in other habitats as well as with communications networks and medical data bases on Earth. No matter what the symptoms or affliction, the medical staff can obtain consultation with other physicians through interactive television links and have at their fingertips the amassed and indexed medical knowledge of the world through computer links. Given basic data from the diagnostic equipment at hand, the computer data bases can provide analysis as well as diagnosis, prognosis, and suggested avenues of treatment.

Surgical facilities located in centrifuged modules of large habitats utilize techniques and procedures that differ little from those used on Earth. However, zero-g surgery is a highly specialized field that requires unique methods and equipment. Some surgery is easier to perform in weightlessness. But surgeons and surgical nurses and technicians must take special training for weightless procedures.

Postoperative, recuperative, and other hospital facilities requiring a high level of asepsis are possible in space habitats. This will lead to a reduced use of wide-spectrum antibiotics, which often have serious side effects.

Summary

When all things are considered and even the high level of potential hazard in space is taken into account, living and working in space you'll be in one of the healthiest and safest of all environments in which you could find yourself. Everyone must of necessity practice a high degree of personal hygiene. Public hygiene is also of a high level because of the requirements to maintain a clean, operative life-support system. Every flight crew and nearly all persons living and working in space habitats are trained in first aid and there are emergency first-aid kits of various types well distributed in spacecraft and space habitats. The excellent communications and computer networks available in space mean that medical assistance and complete medical libraries are quickly at hand. Because of the remoteness of most habitats in terms of the time necessary to return a sick or injured person to Earth, medical facilities in space are among the very best available anywhere. Space is not only a dangerous place to live, but it is at the same time the healthiest.

Figure 14-1.

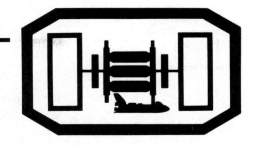

14 Space Habitats

Think of a space habitat as the equivalent of a village, town, or city. These collections of structures and people on Earth exist for multiple purposes, are different because of their locations, purposes, and functions, and have individuality.

Military habitats in space should be thought of as extensions of their counterparts on Earth. A great many earthbound military forts evolved into towns and cities. Most cities in Europe and Asia are located at defensible sites along trade routes where castles or other fortifications were originally built. Many American cities originated from strategically located military outposts located to protect local inhabitants and those traveling past. They proclaim that heritage with city names that begin with "Fort"—Fort Wayne, Fort Worth, Fort Pierce, Fort Myers, Fort Dodge, Fort Collins, Fort Payne, Fort Smith, Fort Lauderdale, Fort Lee, and so on.

However, the space habitats in which you're most likely to be living and working will be those devoted to scientific or commercial purposes and will therefore have a different organization and social life—just as military bases adjacent to towns and cities are definitely different from their civilian neighbors.

Types of Habitats

There are many different habitats. Some may be old, others new. Some may be primitive, while others may offer a life-style that many space pioneers would

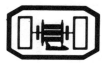

Inside the Skylab, the first American space-station habitat *(NASA)*

consider sybaritic and luxurious. Some may be small and intimate; others quite large, with a definite metropolitan feeling about them. Many may be international in nature, although they may be nothing more than connected modules housing different nationalities, just as some earthly cities have ethnic districts. Some may be "pan-national" in nature, their existence the result of activities of multinational corporations, consortiums, and joint ventures.

Early habitats didn't have to offer a comfortable, almost terrestrial life-style in order to entice people to come there to work and live. Most of the inhabitants of early space facilities were highly motivated to get into space to conduct scientific research, industrial experimentation and development, or commercial ventures. Other than requiring a habitat to provide them with basic life support, the pioneers didn't demand comfort. They're far more interested in safety, reliability, and functional characteristics.

However, as space habitation progresses, we'll see a continuing development of unique life-styles in a series of evolving and improving habitats. There are many problems for habitat designers to solve and many unknown factors that space engineers must learn to contend with. As progress is made in these areas, space habitats will improve as good places to live.

226

Early Habitat Development

The early space habitats such as Skylab were nothing more than launch rocket tanks or stages that had been converted to pressurized living modules. Some of the original planning for such habitats—based on the principle of "it's there, so let's use it"—involved the so-called wet habitat, where the basic enclosure was launched as a propellant tank full of fuel or oxidizer and, once in orbit, fitted out as a pressurized habitat. However, the complexity of cleaning, passivating, and remodeling such a spent rocket stage was felt to be beyond the state of the art at that time.

Numerous NASA and private studies were made concerning the possibility of using the huge Shuttle Expendable Tank (ET) as the foundation for an inexpensive early space station and habitat. The ET is 27.5 feet in diameter and 154 feet long, with an internal volume of nearly 70,000 cubic feet—many times what was available in Skylab. To prevent an accumulation of expended ETs in orbit that might create collision hazards at a later date, ETs were designed to be jettisoned from the Orbiter at a velocity just shy of that needed to achieve orbit and thus enter the atmosphere and break up over the oceans. The ET was only a thin-walled aluminum shell and by far the most inexpensive part of the Space Shuttle flight hardware, comparable to the expendable long-range fuel tanks carried under military fighter aircraft.

The economies that might have been gained by utilizing the ET as the pressurized shell of a space habitat were generally shown by the private studies to be attractive. However, NASA studies indicated far too many technical unknowns were involved. These could have resulted in budget overruns which in the long haul would have made such an ET habitat more expensive and less useful than a dedicated facility such as a space station.

As a result, nearly all habitats have been either launched fully outfitted as space stations, have been assembled in orbit from modules that were carried up

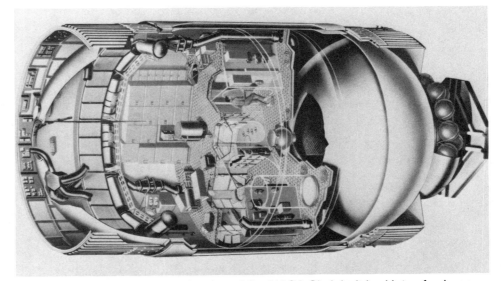

Figure 14-2. Artist's cutaway drawing of the NASA Skylab. It had lots of volume for three people. *(NASA)*

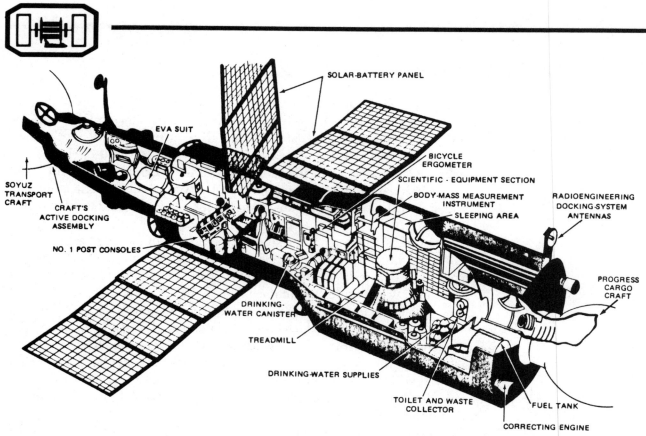

Figure 14-3. Cutaway of the Soviet Salyut space station *(NASA)*

ready for use, or have actually been built in orbit from discrete pieces and parts.

The American Skylab and the Soviet Salyut space stations were of this first sort. Although Skylab was a remodeled third stage of the *Saturn 5* Moon rocket known as the *Saturn S-IVB*, the Soviet Salyuts were designed and built on Earth as space-station units and launched into orbit atop the Soviet *SL-13* "Proton" rocket vehicle.

Studies were made and proposals put forth to utilize as "free flying modules" the SpaceLab units normally carried into orbit and back within the cargo bay of the Space Shuttle Orbiter. This principle was used in the development of the space station in which components of the station would be transported into orbit in the cargo bays of a number of Shuttle Orbiters and then linked together to produce a space station.

One of the problems encountered early in the development of space stations and habitats was the tendency to conduct study after study as space transportation hardware changed and became operational. The large number of studies—at least fourteen space-station studies were made by and for NASA between 1967 and 1980—was necessary because government funding was behind early space activities, and the requests for and appropriation of tax dollars had to have justification at least as strong as the social programs competing for the same tax dollars. There was little understanding that such space-station programs could be, in effect, massive government-supported jobs programs in the high-technology area as well as being, in the long haul, "self-liquidating" programs where primary and secondary spending paybacks more than balanced the funding outlays. The General Accounting Office of the United States estimated in 1982 that every dollar invested in the space program returned seven dollars to the gross national product.

It's relatively easy—albeit more expensive from the standpoint of space transportation costs—to construct a space-habitat module on Earth, fit it out completely before launch, and then carry it into orbit for assembly to a larger conglomeration of modules making up a space station or habitat. This is why most habitats in low Earth orbit and even in geosynchronous Earth orbit will continue to be of this preassembled "dry" modular type.

Right now, it's easier to dock modules together in orbit than to engage in actual construction in the same way that a high rise or skyscraper is put together on Earth by "high iron" workers. In time, however, space activities will require the space equivalent of high-iron workers skilled in assembling space structures either from parts lifted up from Earth or from assemblies fabricated in space from extraterrestrial materials.

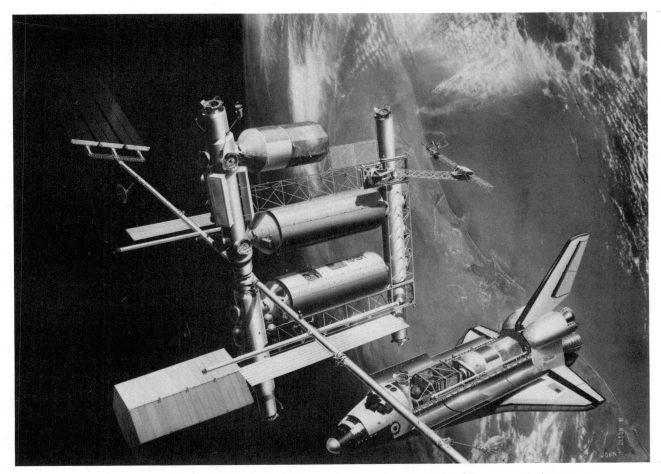

Figure 14-4. An artist's conception of an early NASA space station *(Boeing Company)*

229

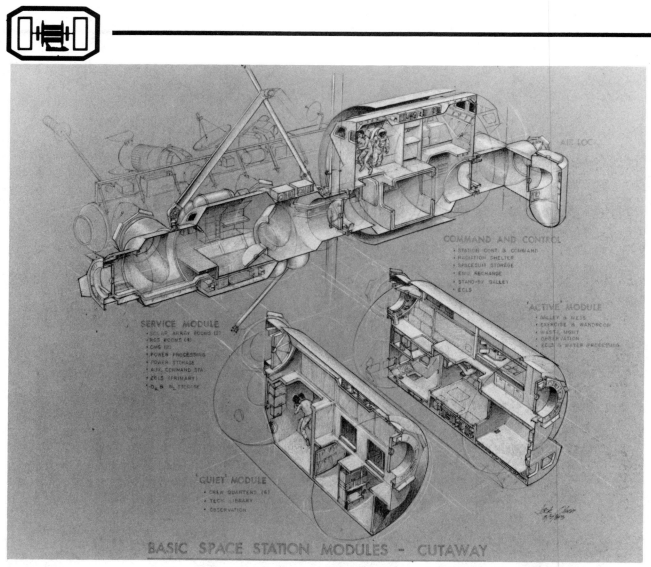

Figure 14-5. Cutaway of an advanced space-habitat module Radiation shelter is part of command and control module *(Boeing Company)*

The Two Basic Habitat Types

There are two basic types of space habitats: those in which all or part of the structure is rotated to produce the pseudogravity of centrifugal force, and those in which all activities and operations are conducted in weightlessness.

Rotating Habitats

Habitats with rotating modules are far more difficult to design, build, and maintain. Therefore, they can be considered to be habitats in which the nature of the life-support system, the stay times of the people involved in the habitat's activities, or all or part of the particular purpose, function, or industrial process of the habitat, requires the pseudogravity of centrifugal force.

230

Some advanced closed-cycle life-support systems may use plant life to recycle carbon dioxide to breathable oxygen by means of the photosynthesis process as well as to provide supplementary food for the inhabitants. Plants exhibit a strong geotropism, the inherent tendency to send roots down and stalks up. They need gravity and in a weightless condition exhibit reduced growth, delayed growth, and inability to blossom and fructify to produce seeds, kernels, and other reproductive elements, many of which are major foods for human beings. Many experiments involving astroagriculture have already been made.

Some industrial processes will require a pseudogravity force because they depend upon density differences for operation or utilize liquids that must be kept contained.

Either the distances, propellant requirements, locations, or other physical factors involved may make it far too expensive to change habitat crews regularly. Perhaps the industrial process itself will require people to stay with it to operate or monitor it for long periods of time.

Such requirements for long stays in space may make it necessary to centrifuge at least the living quarters to prevent any physiological deterioration that may result from long-term living in weightlessness.

The calcium-resorption problem or the atrophy of the cardiovascular system may make it necessary to provide pseudogravity for long stays in space, or

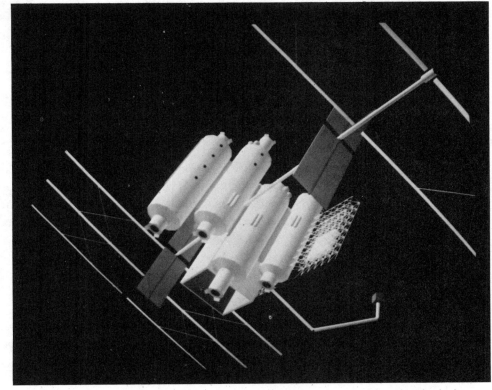

Figure 14-6. Computer-generated image of a future space habitat that includes a hangar for servicing spacecraft *(Boeing Company)*

Figure 14-7. Military habitats won't operate differently from civilian ones, but they'll have much different uses. *(Art by Sternbach)*

it may turn out that a large percentage of people suffer so badly from weight-less-induced motion sickness that they can't adapt to zero-g for days or even weeks or perhaps ever.

In a centrifuged habitat, you'll think you're aboard ship somewhere on Earth's oceans, in an Arctic pipeline station, or a smaller version of New York's Grand Central Terminal without so many people. There will be floors, ceilings, walls, doors, tables, chairs, benches, food that doesn't float, liquids that don't get out of control and go everywhere, and all the ordinary comforts of home.

However, there will be some important differences as well.

The Coriolis-Force Problem

The centrifuged habitats will be *large* in comparison to the weightless ones because of your reactions to something you've seldom encountered on Earth, something that therefore is understood by very few people.

This factor is *Coriolis force*, named after the French mathematician Gaspard Gustave de Coriolis (1792–1843), who first described it.

It's a consequence of Newton's First Law of Motion: "An object at rest will

The "Pilgrim Observer" was a 1969 concept of a rotating space habitat that could also be used for planetary exploration. Its three rotating arms contained a nuclear power station, a crew living area, and a garden or "space farm" to recycle the oxygen in the habitat's atmosphere. It was available as a plastic kit and was one of the most predictive of all early space kits. *(G. Harry Stine)*

remain at rest, and an object in motion will remain in motion in a straight line, until acted upon by an external force."

A good illustration of Coriolis is shown in Figure 14–8, which represents a stereo turntable as you look down upon the rotating record. Using a piece of chalk, draw a "straight" line from the center to the outer edge of the turntable. To you, an outside observer, the chalk appeared to travel in a straight line from the spindle to the edge. But stop the turntable and look at the chalk mark on the record. It's a line that spirals outward from the center spindle.

If you'd been riding around in the record turntable when the chalk mark had been made, it would appear that some "force" deflected the chalk from a straight line and made it curve. The "force" is not really a force at all, although it's called Coriolis force.

Coriolis force makes the prevailing winds on Earth blow from the west in the northern hemisphere and from the east in the southern hemisphere instead of north and south as they would on a nonrotating Earth.

233

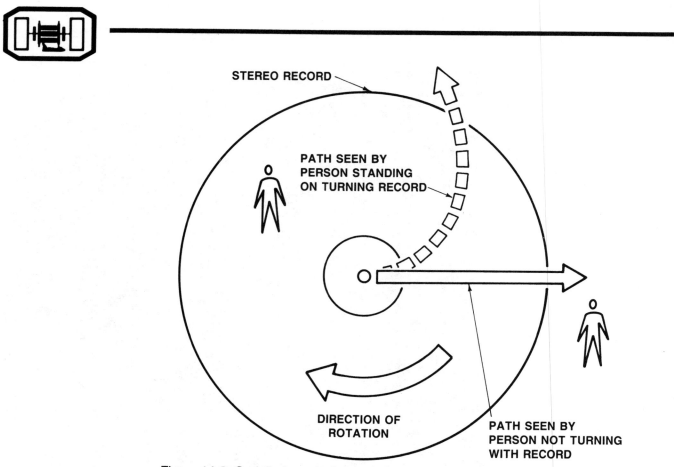

STEREO RECORD

PATH SEEN BY
PERSON STANDING
ON TURNING RECORD

DIRECTION OF
ROTATION

PATH SEEN BY
PERSON NOT TURNING
WITH RECORD

Figure 14-8. Coriolis force isn't a real "force," only an apparent one. *(Art by Sternbach)*

How will this affect you in a rotating habitat?

Look at Figure 14–10, which represents a rotating habitat shaped like a doughnut and called a "torus." As it rotates, it creates centrifugal pseudogravity force that holds the tiny human figure on its inside surface.

Suppose the torus is 1,000 feet in diameter and rotating at 1.71 revolutions per minute to create one g on its inside surface. If you're the tiny human figure and if you jump off a stool 21.5 inches high, Coriolis force makes you land about 2 inches to the side of where you aim.

Coriolis force exerts strange effects upon people, too. In a habitat rotating at several rpm, simple movements become complex, and there are strong visual illusions. A rapid turning of your head can make stationary objects appear to gyrate and continue to move once you've stopped moving your head. Coriolis force also creates cross-coupled accelerations of the fluid in your inner ear, your balance and equilibrium organ. This in turn can lead to severe motion sickness. You can probably adapt to these in habitats rotating as fast as 3 rpm, but it may take some time.

Human Tolerances to Rotation and Gravity Gradient

There are two other reasons why centrifuged habitats are very large.

The first of these has to do with your tolerance to rotation. You've experi-

Figure 14-9. Rotating habitats will have be very large to prevent the problems caused by Coriolis force. *(NASA)*

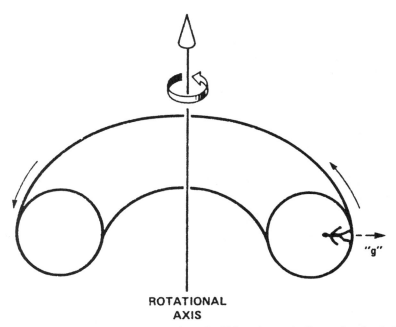

ROTATIONAL AXIS

Figure 14-10. In a rotating habitat, your head will be closer to the axis of rotation than your feet and will therefore be subjected to less pseudogravity created by centrifugal force. *(NASA)*

235

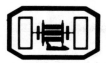

enced dizziness caused by rapid rotation, so this factor isn't anything new to you. Again, this is a human factor that was first encountered in aerobatic aircraft and, later, in uncontrolled rotations experienced by parachutists. The U.S. Air Force human-factors researchers encountered the human-tolerances limits to rotation when testing the first jet aircraft ejection seats, some of which would tumble rapidly through the air after coming out of the airplane and cause the occupant to lose consciousness. So they set about to learn what the tolerances were. Some of the early standards were somewhat arbitrary. The Air Force was persuaded to increase their tolerance standards when a research group, of which I was a member, measured the rotation of an ice skater doing a spin and discovered that it far exceeded the standards.

Most of the research on human tolerance to rotation has therefore come from Air Force testing. There's an upper limit to how fast you can be spun before you become dizzy or, in human-factors terminology, lose orientation. Although people can function at rotation rates as high as 10 rpm, such a high spin rate causes them to suffer from a high degree of disorientation created by Coriolis forces. Spin rates of 3 rpm don't.

The second has to do with a Coriolis-like effect that may be a benefit to some space industrial processes but could be a problem for you. It is called *gravity gradient*.

Look at Figure 14–10. It's not drawn to scale but has been exaggerated for the sake of explanation and understanding. Suppose you're the human figure shown standing on the inside surface of the rotating habitat. The centrifugal force created by rotation depends upon the distance of the affected mass from the center or axis of rotation. The farther from the axis, the greater the centrifugal force. The centrifugal pseudogravity force at the floor level is, by design, one g. But, if you're 6 feet tall, your head is that much closer to the axis of rotation and therefore is subjected to *less* centrifugal pseudogravity force. If you're 100 feet from the axis, the centrifugal force on your head is 6 percent *less* than that at your feet. If you lift something, it appears to become lighter as you increase its distance from the floor and decrease its distance from the axis of rotation.

There's also a gravity gradient that can be created by Coriolis force. Suppose you start walking around the inner circumference of your little 100-foot-diameter rotating habitat. If you walk in the direction of rotation, you will be rotating faster than the habitat and therefore there will be more centrifugal force holding your feet to the floor; you'll seem to weigh more by a factor of as much as 20 percent. If you walk in the opposite direction *against* the direction of rotation, you'll seem to weight 20 percent less.

To an Earth person, these enormous variations in the magnitude and direction of the apparent gravity force can be unexpected, and are certainly highly unusual. They may therefore cause an individual to become quite ill and suffer visual hallucinations. It will probably take weeks for the person to adapt to these effects, if he happens to be an adaptable person; if not, he'll have to go home to Earth or to a weightless habitat. Space sickness, once a fictional concept of science-fiction writers, is a reality. It can be debilitating, and therefore extremely serious, in a hostile environment such as space.

Centrifugal Force

Coriolis forces, gravity gradients, and problems with human tolerance to rotation rates become less important and have fewer physiological and psychological effects as the diameter of the space habitat increases. This is why all centrifuged space habitats must be very large and why most small habitats have to be of the weightless type.

The basic equation describing the magnitude of centrifugal force is:

$$f = mv^2r$$

where f = centrifugal force, m = mass, v = angular velocity in degrees per second, and r = the radius in feet of the rotating habitat.

Therefore, a smaller habitat can be tolerated if the required pseudogravity g-force can be reduced. Figure 14–11 shows the habitat rotational rate in rpm necessary to produce various g-forces for various habitat sizes.

If you want to live in a comfortable one-g environment in space, you're going to have to select a very large habitat, which means it will have lots of people living in it. It will also mean that you may have to wait longer to live in space, because the big habitats are more expensive, take longer to build, and therefore aren't as numerous.

Satellite Rotational Rate Necessary to Produce Various G-Forces for Various Satellite Radii

(Rotational rate expressed in rpms)

Radius (feet)	G-force 1.0g	0.5g	0.25g	0.1g
5	24.17	17.09	12.08	7.64
10	17.09	12.08	8.54	5.40
15	13.95	9.87	6.98	4.41
20	12.08	8.54	6.04	3.82
25	10.81	7.64	5.40	3.42
30	9.87	6.98	4.93	3.12
40	8.54	6.04	4.27	2.70
50	7.64	5.40	3.82	2.42
60	6.98	4.93	3.49	2.21
80	6.04	4.27	3.02	1.91
100	5.40	3.82	2.70	1.71
120	4.93	3.49	2.47	1.56
140	4.57	3.23	2.28	1.44
160	4.27	3.02	2.14	1.35
180	4.03	2.85	2.01	1.27
200	3.82	2.70	1.91	1.21
300	3.12	2.21	1.56	0.99
400	2.70	1.91	1.35	0.85
1000	1.71	1.18	0.85	0.54

Figure 14-11.

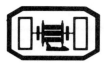

If you can tolerate 0.1 g and if subsequent medical data from long-duration space living indicates that such a reduced-gravity environment improves the physiological problems associated with prolonged stays in weightlessness, you can live in a 200-foot-diameter habitat, *if* you can manage to adapt to the strong Coriolis and gravity gradient effects present in the smaller habitat.

Transition Problems

There's also another problem involved with living in a centrifuged habitat, particularly if your work requires that you make the transition from pseudogravity to weightlessness.

You'll enter and leave the rotating centrifuged module through a complex compartment in the hub of the module. This is the place in a rotating module where there is the least difference between the speeds of the rotating part and the nonrotating part. Depending upon the size of the rotating module, you'll get to the hub along a structural spoke of the "wheel" by climbing a ladder or a stairway, or by riding "up" to the hub on a personnel conveyor or an elevator. As you go from the rim of the module to the hub, the pseudogravity force becomes less until, at the hub, you are effectively in the microgravity of near weight-lessness—so close to being totally weightless that you can't perceive the minute difference. Returning to the rotating module and pseudogravity means that you'll reverse this procedure.

You may have trouble adjusting your equilibrium senses from the rotational world of pseudogravity to weightlessness and back again with such rapidity. Some people will suffer from motion sickness because of this. Although most people are highly adaptable, some individuals will never be able to adapt to the change—they can live in perfect comfort in a 0.1 g centrifuged module, for example, but cannot tolerate a daily transition to and from weightlessness. At the present time, there's no reliable test that you can take on Earth that will indicate whether or not you'll be able to adapt quickly, over a period of several days, or never. Although one may be developed as a result of accumulating data, it appears that you'll just have to go there to find out.

Weightless Habitats

The habitats that exist continuously in weightlessness will have different problems but are, by and large, far more fun and interesting to live in if you're highly motivated to go into space and live there.

Weightless habitats can be designed and built much differently from those with centrifuged modules. They are far less complex because they don't have to withstand the structural stresses created by a rotating module, nor are they affected by the enormous gyroscopic effects of large-diameter rotating portions. This last can create huge problems with attitude control of the habitat to maintain its orientation with the Earth or the sun.

238

Weightless habitats can also be designed and built far smaller and more compact than centrifuged habitats. Here, Coriolis forces are no longer design and living factors because they don't exist. And every cubic inch of volume inside a weightless habitat can be used.

Living conditions in the weightless habitats will be as radically different from those in centrifuged habitats as they are from Earth conditions.

Until more data are gathered concerning the long-term physiological effects of weightlessness on people, a weightless habitat won't be a permanent home for you. It'll be manned by transient crews rotated back to Earth every six months or so. In this regard, these weightless habitats will resemble most of the offshore oil rigs on Earth where crews are rotated on a regular schedule.

The analogy to offshore oil rigs also holds when it comes to supplying these habitats because their life-support systems may be of the open type requiring regular replenishment of oxygen, nitrogen, lithium hydride canisters, and activated-charcoal odor-control filters. Excess water created by human metabolic processes will also have to be removed because it may not be possible to vent it overboard to space; such an action may contaminate the surrounding space environment, which in turn may be at the heart of the particular industrial process for which the habitat exists. And somebody is going to have to collect the garbage and other waste for removal to Earth.

The biggest problem with weightless habitats lies in the minds of their engineering designers.

Very few people yet have the capability to truly think in three dimensions because, as land animals, they've evolved in a two-dimensional world and often experience difficulty while operating in a three-dimensional one. This is one reason why on Earth there are so many automobiles and so few airplanes. It's far easier to operate an automobile, in two dimensions. It takes a great deal more training and almost constant practice successfully and safely to operate an airplane in its three-dimensional element.

This "dimensional" shortcoming affects many of those who design and lay out weightless habitats. The shortcoming will disappear within a decade as more and more people learn how to live in weightlessness and therefore how to *think* in the three-dimensional terms of weightlessness. After the first children are born in weightlessness and grow up to design future habitats, the transition will be completed, and living quarters will then be truly designed and built for weightless living.

Designers who specialize in marine vessel design and in that almost extinct field of designing railway sleeping cars or "Pullmans" have come closest to knowing how to design for volume rather than floor space. But most designers simply don't totally understand how to design for weightless living. While this was anticipated and could be tolerated in some of the pioneering space habitats such as Skylab and Salyut, it was a wasteful shortcoming that revealed earthbound thought patterns.

It's amusing now to look at the basic shortcomings in design that were made because of engineering conservatism coupled with earthbound thought processes.

For example, the designers of the first NASA Space Shuttle Orbiter derived its basic layout from the airliners of the day. The Orbiter had a "flight deck" and a "mid-deck." Both were laid out as though the Orbiter were an airliner flying horizontally through the Earth's gravity field.

None of the designers considered the basic problem of "crew ingress" (entry) with the Orbiter in a vertical orientation on the launchpad, and a number of quick fixes had to be introduced at the last moment to ensure that the flight crew and passengers could get into their seats, which, in the launch orientation, were no longer on the floor but on the wall.

Except for the hatches on the roof of the flight deck for the ejection seats of the early *Columbia* flights, there's only one small hatch forty inches in diameter located on the left side of the mid-deck through which the flight crew can enter (ingress) and exit (egress) the Orbiter under normal circumstances *and* emergencies. There's another forty-inch-diameter hatch leading into the airlock and payload bay on the aft bulkhead of the mid-deck. Theoretically, it's not possible to put anything into the Orbiter's crew module if it exceeds this forty-inch limitation.

Although the Space Shuttle flight crew floats in weightlessness in orbit, they were provided with ladders for interdeck movement as well as a stepping stool so that small people could "stand" up to the payload control panels located on the aft bulkhead of the flight deck.

Even some of the early designs of the space station revealed this earth-bound orientation in thinking.

Living in the Weightless Habitat

The best description of living in a weightless habitat is that it's like living underwater without the water.

Living and working quarters in weightlessness are quite different from anything that you'll encounter anywhere else because the quarters use *cubage* rather than just floor and wall space.

Even the most primitive weightless habitats won't use the "dormitory" or "barracks" technique of providing living quarters. Individual compartments are used to a far greater extent. Among the reasons why individual living quarters or "rooms" are used are both safety and structural strength.

Dividing the large internal volume of a habitat into smaller volumes with bulkheads permits structural loads and forces to be routed through multiple stress paths in the same manner as the pressurized hull of a jet airliner is designed with multiple load paths. Although there are design trade-offs that engineers make between single-wall self-supporting monocoque structures and multicellular enclosures, the internal bulkheads of smaller personal compartments help distribute the loads and, with only a small mass penalty, also provide pressure integrity and safety similar to the watertight compartments of a large ocean vessel. If pressure is lost in a single compartment because, say, of penetration by that rare meteor large enough to broach the meteor shield, the affected personal compartment can be sealed off quickly to prevent the entire internal atmosphere of the habitat from leaking out into space.

Figure 14-12. Various types of restraints useful for moving or staying in one place under weightless conditions *(NASA)*

Another important reason for building habitats with personal living quarters involves human comfort, preference, and need for privacy. These factors are behind the way we build our dwellings on Earth. Although primitive and poor cultures still use dwellings in which a single room is lived in by many people, as quickly as their means permit it, these people rapidly build new dwellings or add onto existing dwellings so that more and more personal privacy becomes available.

Therefore, in most of the habitats in which you'll be living, you'll probably have the choice of a single living compartment or a "double" which is shared with someone else.

Although the way your personal living space is designed may at first seem strange, you'll quickly grow used to it and realize that a great deal of careful thought and consideration went into it. At first, to your earthbound way of

241

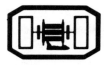

thinking it may seem far too small, but remember that the walls, floor, ceiling, and every little nook and cranny can be used in the weightless environment. There's no furniture as you'd normally think of furniture, because those earthly artifacts are primarily designed to *support* a human being against the force of gravity. Zero-g furniture serves different functions, some of which on Earth are only secondary ones.

There's no bed as you've known it, but you'll have your choice of sleeping in a sleep sack or under the gentle restraint of belts, either of which will keep you from floating around the compartment in your sleep and possibly banging into something that could cause injury to you.

There are no chairs because those are earthly things useful only in a gravity field.

There's a working desk that folds down out of the wall or bulkhead. It may have knee or leg restraints "underneath" so you can tuck yourself up to it for working. Although this desk is similar to the meal tables in the common room or wardroom of the habitat, you won't be eating in your compartment. You might use your desk for writing, although that primitive form of communication and data keeping will be obsolete because you'll have a computer terminal and keypad on or in the desk, and you'll either have access to the habitat's general-purpose computer with its network links to data banks, or you'll have your own little dedicated personal computer, which can also be networked. The working desk will also contain your communications terminal—your telephone, so to speak.

The compartment will have its own entertainment center capable of receiving television from Earth or from a video player, a stereo music system, and a computer gaming system. Some of this may be incorporated into your computer terminal.

There will be storage lockers for personal items such as clothing, books, personal-hygiene items, etc.

Some habitats will have "room with bath," and some will have a separate general-use bathroom. Others will contain only a simple wash-up station. The size, complexity, and age of the habitat will determine how the personal-hygiene and public-health facilities are arranged and allocated. The newer and larger the habitat, the more sybaritic the facilities will be. Expect the primitive, however, and then you won't be disappointed. Space living is still not as luxurious as living on Earth, but it's far and away better than what your forefathers had to endure on any terrestrial frontier in the last two hundred years.

You may or may not have a personal view port in a bulkhead of your compartment. It depends again on the size of the habitat and how recently it was built. Some habitats have view ports only in common rooms or wardrooms. But all habitats have view ports, windows you can look out of, in spite of the fact that such view ports can never be truly sealed tightly and are the cause of much atmospheric leakage. However, view ports are part of the human side of space living. The Skylab astronauts discovered that they spent a great deal of their free time in a very simple recreation: looking out the port in the Skylab wardroom at the Earth as they passed over it at an altitude of more than 200 miles. All habitats since have had view ports, because people simply like to look out the window and watch the world go by.

In this regard, living in space isn't that much different from living on Earth. There are differences, to be sure. And, in the same vein, living with other people in space is at once similar to and different from how it's done on Earth.

Figure 15-1. Working in a space factory *(Art by Sternbach)*

15 Social Aspects of Space Living

Before people began to live and work in space, early space planners spent an enormous amount of time and effort attempting to determine many of the social aspects of the space colonies whose development they advocated. Although space colonies were basically technically feasible, most of this work was far in advance of the willingness of any government, corporation, or other large funding group to create space colonies.

Space-colonization social studies tended to be strongly utopian in nature. There's nothing wrong with this; it gives people something to work for in terms of long-range goals. Every generation develops its own concept of utopian living, and in the last few centuries many groups of people have managed to achieve their utopias. In so doing, however, they discovered that the real thing turns out to be a far cry from what they had envisioned when they started. They also discovered that the achievement of a utopia requires an unprecedented, unanticipated amount of hard physical and mental work accompanied by worry, anxiety, panic, controversy, arguments, tragedy, accidents, and far more deaths than any of them would have accepted at the start.

This basic description applies to any and all human frontiers—past, present, and future.

Prudence would dictate that we should carefully study the lessons learned by our forefathers on the old frontiers. But this doesn't seem to happen. Each person and group learns what it thought would be applicable, forgets the rest,

245

forges ahead with a biased viewpoint of frontiermanship based on legends, folk tales, fiction, television shows, and wishful thinking. It thus proceeds to make far too many mistakes that work to increase the monetary and human costs of the new frontier.

Space is no different, even though space is an opportunity to start afresh without the ghosts of past mistakes to haunt us, an even better opportunity than our forefathers had when they immigrated to North America.

The Primitive Social Sciences

Early space advocates attempted to tackle many of the social aspects of space colonization with the meager and inadequate tools of the "social sciences" of the time.

These adjectives are deliberately used here because the social sciences have yet to reach the point where they're predictive sciences such as physics, chemistry, and astronomy. At some future time, perhaps because of what can be learned from groups of people living in the relative isolation of space habitats where social experiments can be conducted, the social sciences will benefit enormously from space and make their final transition to predictive bodies of knowledge.

But this point hasn't been reached yet, and it's extremely difficult to get social scientists to admit this. They'll argue loudly in their cause. But they truly cannot yet predict the future behavior of social systems or an economic unit as large as the United States of America. If they could, there would be no cyclic economic problems of the boom-and-bust sort, and a nation's economy could be truly "managed."

Only one group of people has actually managed to design a large social organization that turned out to be capable of considerable growth and that has lasted more than two centuries. They were fifty-five men who were not social scientists but military officers, financiers, public officials, merchants, lawyers, and planters who met from May to September 1787 in Philadelphia to thrash out a Constitution for the United States of America. And for the next two centuries, the people of the United States, who weren't social scientists either, spent a lot of time and effort making that Constitution work. Even after two centuries, it isn't a perfect document or form of government, and it requires a high degree of education and a lot of work to continue to make it work. It is, as Dr. Daniel Boorstin points out in his book *The Republic of Technology*, an experimental document to establish an experimental institution that always changes.

Armed with the best tools of the social sciences, however, some space advocates actually attempted to "design" populations for space colonies and to work out criteria for selection of those who'd be permitted to live there. Others attempted to design the social institutions of space colonies—governmental forms, codes of law, economic systems, monetary systems.

Far too many of these attempts were made using the easy way out—a centralized, authoritarian, bureaucratic, and totalitarian political and govern-

246

mental scheme—because this is the simplest approach. It is not, however, the most workable.

Other planners went to the opposite extreme and designed almost anarchistic "let it all hang out" romantic communal-type institutions based upon the recent twentieth-century romantic movement that began in the late 1960s and occupied most of the 1970s.

Neither approach will necessarily work but for reasons usually not fully appreciated by the space-colony planners.

The institutions of space may eventually evolve beyond the institutions of Earth after which they were patterned. But they'll all start, even as the United States of America did, with a basic foundation of past and existing institutions that people *know* will work, however poorly.

Institutions

At this point, I should define the word "institution" because many readers use the word in the context of their own experience and education rather than the context in which it's used in this handbook.

An institution is a group of people. It's organized for some purpose or purposes. It follows rules agreed upon by the people in the group. It has a structure that, in its simplest form, amounts to a leader and a few followers. It has internal protocols so that people know who's in charge and know the relative ranking of individuals within the group organizational structure. These protocols include rewards for activities that benefit the group as well as punishments for breaking the rules. In line with this, the group may develop a means of keeping score so that those who are rewarded may be recognized commensurate with their activities and those who must be punished may be suitably dealt with. There are procedures for admitting new people to the institution and for group members to leave voluntarily or forcibly. While these elements are concerned with internal politics, the institution also has some rules that guide its relations with other institutions.

It's important to realize that an institution is a group of *people* and not a collection of ideas, rules, practices, or customs.

Culture

An institution also is not "culture," a term with about as many definitions as there are social scientists. For our purposes, culture is the way in which individual people live in or with the institution. Culture is transmitted from generation to generation by learning and the educational process. It includes the relationships between individuals and the institution, the way they work within an institution in handling matter and energy, and the way they operate in handling symbols such as speech, music, the visual arts, and the human body itself. Culture is the sum total of the things that people do as a result of being so taught.

247

Institutions and Culture in Space

In space, you'll find no space culture per se because it hasn't had time to develop yet. It takes generations for a separate culture to evolve. In the meantime, you'll take your earthly culture with you and adapt it to space. You'll teach newcomers and children what you've learned that works and doesn't work. You'll develop the space culture yourself.

But you'll find institutions in space because people *must* organize themselves in this new environment that is somewhat hostile and deadly to the rugged individualist. Actually, this last is also true of nearly every place on Earth. The places where the rugged individualist can exist in total isolation, much less live well, have been disappearing for centuries as a modern technological life-style requires the cooperation of a large number of people in a variety of institutions.

Although your living style in space will be adequate and comfortable, don't expect to live in any habitat in your concept of a utopian institution. Utopias don't exist in space yet. Most habitats have been sponsored, paid for, built, and operated by public or quasi-public institutions with a bureaucratic structure. Most habitat organizations are pragmatic copies of similar ones on Earth.

Paramilitary Habitat Organizations

The early and small habitats are most likely to be quasi-military, authoritarian hierarchies that have derived from the paramilitary nature of the early space programs of both the United States and the Soviet Union. Although overtly

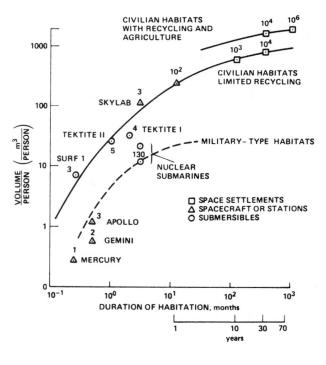

Figure 15-2. Chart showing personal space and volume required by people (NASA)

248

"civilian" in nature, these have been staffed and managed by people who were trained in the military services or who had spent most of their lives working for such services. All early space vehicles were modified military rockets. A large percentage of the astronauts and cosmonauts had previously been military aircraft pilots. (In fact, in most of the world today, the fact that a person is an aircraft pilot automatically means that he or she was trained in the military forces of the particular country; and "once a soldier, always a soldier.")

This shouldn't necessarily be considered a bad way to run things early in the space-habitation business. Naval forces of the world over many centuries have developed highly successful maritime institutions relating to the adequate performance, safety, health, and comfort of large groups of highly trained people living and working in isolated, dangerous vessels, habitats, and environments. Most of the data upon which the design of space habitats is based comes from naval and aeronautical research in the field of human factors, including the psychology of small groups living closely together in isolated, dangerous places for long periods of time.

Technobureaucratic Habitat Organizations

You may also find yourself in a habitat that runs on a system of technocratic centralized control somewhat similar to the paramilitary organization just discussed but far more highly concentrated in the technical area because of the purposes of the habitat. This is especially the case if the habitat happens to be a construction base or a research-and-development laboratory. Again, this institutional form of organization and governance isn't necessarily bad or restrictive.

If the habitat is primarily under the control of a large domestic or multinational corporation or a consortium of corporations, you'll most likely find an organization based on the bureaucratic management techniques of the parent earthly organization. This may turn out to be a participative meritocracy.

Participative Democratic Habitat Organizations

The least likely form of social organization and government you'll find in space habitats will be the self-organized popular participative democracy operating by town meetings. Such an organizational form requires either elected leadership of volunteers who must govern in their spare time away from their everyday work in the habitat, or the employment of specialized hired habitat managers, which naturally shades over into the bureaucratic organizational types discussed immediately above.

The Critical Factor

Under no circumstances should you design, develop, or participate in any institution in which any individual member is not responsible to the group or to

249

the group's leaders, and in which the group's leader isn't responsible to either the group or some higher and accessible authority for any action or activity of either a committed or omitted nature. Eschew absolutism.

Absolute power does not corrupt absolutely; absolute immunity does.

Emerging Space Institutions

No institution remains static and survives. Nor can an institution designed for one set of circumstances be expected to work successfully in widely different circumstances. This is especially true of the emerging social institutions of space habitats. While the social organizations of space living may start out taking one of the forms discussed above, they'll evolve as you and your colleagues develop your own sense of community and learn how best to organize for your personal and collective survival in a new environment.

You'll have to make some very difficult personal decisions, most of which don't have to be faced by citizens of the modern superindustrial culture of the United States. Most of these will involve assigning priorities to personal freedoms. The most important element of our evolving social institutions on Earth seems to be the preservation of the personal freedom of choice consistent with and best contributing to the survival of the institution. This may be and has been looked upon as a sort of implied acceptance of a social contract. Probably the most important of all freedoms of choice to be preserved in any system is the personal freedom to leave the system in order to seek another system that may offer you more of what you desire.

The Great Social Experiment of Space

The evolution and development of the various space institutions and cultures will be fascinating both to observe and to participate in, whether or not you're a social scientist. Certainly, this ongoing and growing process will provide the social scientists with some of the best hard data they've had in a long time—if they'll pay attention to it—as well as add rigor to their specialties.

As continued development of space leads to more and more settlement and more and larger habitats (which will finally automatically be called "colonies"), you'll begin to form more and more institutions as you've already done on Earth. Each of us belongs to several institutions—church, service club, social club, hobby club, professional society, and even local and state governments, as well as national and international organizations. To some extent, you'll take most of your memberships and interests in these institutions into space with you and either adapt them to the evolving culture of space living or form new ones using the old ones as models. You'll certainly see a rising degree of community identity and integration which over the long haul will inevitably lead to a desire on your part to separate your habitat's decision-making powers and governmental organization from those on Earth. You may even see a growing integration of groups of habitats leading to some manner of "inter-space" organizations equivalent to interstate and international groups on Earth.

250

Working together—the same on Earth and in space *(Newport News Ship & Drydock Company)*

Social Revolution in Space

Some scenarios forecast revolutions of the sort in which the thirteen colonies of North America broke away from the English crown, but deem unlikely the sort of revolution that's taken place far more often on Earth, in which the *ancien régime* is replaced by a new ruling group. The sense of cultural integration in a space habitat is far more likely to lead to the separatist type of revolution rather than a move to "throw the rascals out."

However, such considerations won't be serious—or taken seriously—until far more people are living in far larger space habitats with generally closed ecologies. It's difficult if not impossible to revolt against earthly ties if your oxygen, nitrogen, food, and other supplies have to be ferried up from Earth on a regular basis in order for you to survive in a space habitat.

The Space Utopias

There's been a great deal of study, speculation, and outright wishful thinking on the part of space advocates and habitat designers who've attempted to lay out a "perfect" colony or habitat. They have not only done a careful job of selecting the size, shape, internal layout, materials, structural design, and life-support systems, but have also attempted to treat the potential inhabitants as something like human machines in perfect accord with the tenets of materialism.

The screening and selection of people to dwell in space have been seriously considered—often in near-complete ignorance of human nature and the history

251

of analogous settlements on Earth. It has been suggested that people be selected on the basis of intelligence (the intellectual utopia), physical fitness (the superman utopia), psychological fitness, compatibility, reliability, and a host of other parameters, most of which are determined on the basis of somewhat flexible standards or testing results that can be interpreted and therefore biased in any direction the selecting committee really desires.

Obviously, it will cost so much money to put people into space and keep them there that the selection process is necessary to screen out the deadwood. If anybody really knows how to do this, there are thousands of corporations as well as various departments of the United States government that would be eager to embrace this knowledge and pay well for it.

People today seem to forget our own North American history, to say nothing of the history of other parts of the world. Many of our recent progenitors came to America because they *had* to. They were debtors running from bill collectors and prison terms, criminals on the run from the police for various offenses including political heresy, and people who were generally misfits in their own European cultures. The continent of Australia was initially colonized by criminals; it was a penal colony to which you could be banished with little or no chance that you'd ever show up again in England to continue making trouble.

Basic Personal Requirements

Selecting people for space is quite straightforward and probably won't follow any of the utopian principles. The selection will be pragmatic.

The personnel requirements for space workers are not much different from those of terrestrial workers in remote and isolated work sites such as Prudhoe Bay or the offshore oil rigs, to mention but two.

The first requirement is always: "Can you do the job that we need to have someone do out there, and how well have you done similar work in the past?" That's the initial screening process, and its basically technoergonomic in nature. If they're looking for life-support-system maintenance engineers (spacegoing sewage workers), and you're experienced only as a production biotechnician, don't call us, we'll call you.

If you can do what needs to be done, *then* you'll get the inevitable physical examination not only to ensure that you're healthy enough to cast a shadow and to pinpoint any minor physical factors that might affect your ability to do the necessary work in space. The exam will also reveal potential health problems that could affect the future liability of the government agency or company for whom you'll be working.

This will be followed by psychological testing based largely on the successful techniques of "personality profile screening" used by the USAF Strategic Air Command to select bomber and missile-silo crews and by the U.S. Navy to select the crews for missile-launching submarines. These have been extremely effective in preventing psychopaths and other people with personality problems from obtaining any control over nuclear weapons and thus creating a Dr. Strangelove scenario.

The people thus screened and selected won't be superpeople any more than those who work at Prudhoe Bay.

Some will turn out to be misfits in the space environment, but there's no known form or type of examination that is 100 percent effective in this regard.

People like yourself who live and work in space will be healthy, reasonably intelligent, and normally social individuals who are different from one another. They will be willing to make trade-offs between personal freedoms and the general well-being of everyone in the group of people who are separated from the hostile universe only by a thin pressure bulkhead.

This means that there will always be personal problems and internal political problems in any space habitat. Be prepared for them, and also be prepared to solve them, because it's a long way home.

These are a few of the items that you should look into and get information about on your own hook when you're deciding who to go to work for and which habitat you'd prefer to work in. This is not much different from selecting an employer on Earth. All of the suggested check items that follow will have some effect on how you live in space. The priorities you select, the things that are most important to you, and the type of organization and its operating policies that you finally decide to go with are matters of personal choice. You'll select a different habitat and system from the one I would select.

Money in Space

Money is an interesting new aspect of space living. You'll be paid very well for working in space because you'll be doing something new and because the potential hazards are great. But you probably won't be able to spend your paycheck there until the size and number of habitats and colonies grow.

In some habitats, money won't even exist, because of the size of the facility (there's no place to spend money and nothing to buy) or because it's run in a paternalistic manner by a government or corporation that takes care of all of your personal needs while you're in space.

In other facilities, "money" will exist only in computers used to keep track of things, because the habitat is operated much like the corporate towns in the mining districts of Earth. In these "company habitats," your account will be credited regularly with your "salary," and when you "buy" food in the company cafeteria or use the habitat's waste-handling facilities, your account will be debited. It may also be debited on a regular basis for the use of air, water, and electrical power. Whether this is done at all and the extent to which it's done depends mostly on the way the comptrollers and accountants look at things, how the organization's bookkeeping system is set up, or how the government tax collectors and attorneys decide to interpret the then-current tax laws. In such moneyless habitats, barter and in-kind trade become the most common form of value exchange, which is why they can be analogous to a strip-poker game.

In other habitats, you may be issued pseudomoney in the form of chits or tickets for meals, water, baths, air-lock cycles, or a number of other activities that must be kept under control because of costs, logistics, or simply the accounting system used by the habitat's primary organization. This sort of thing is widely used on Earth in a large variety of organizations ranging from schools to international sporting competitions. A ticket or a chit is pseudomoney because, unlike money, once it is used it becomes thereafter valueless. A punched-out meal ticket is a worthless piece of tagboard. However, if it hasn't been used or has been partially used (with some unused punches remaining), it has value just as money does. Tickets and chits themselves become things to trade for value received, and this system in a habitat is analogous to using poker chips in a poker game.

Other habitats, especially large ones, will have a definite form of money. This valuta may take the normal form of specially printed slips of paper or some sort of readily identifiable and difficult-to-duplicate token. However, some *new* form may also appear, because many of the everyday items we take for granted on Earth may be exceedingly rare in space.

Valuable Items in Space

You'll discover that many things have a definitely different value in space. Anything that isn't made in space and that has to be transported to the habitat from Earth will likely be rare and expensive because of the energy required to get it there. On the other hand, many things that are expensive on Earth are relatively commonplace in space and therefore cost much less.

Paper is an example of the first category. Paper must be made on Earth from wood pulp. Gravity and a lot of water are required in the papermaking process as it now exists. At some future date, something resembling paper will probably be manufactured in space, but it will be made from a material other than cellulose fibers. Until that happens, paper will have to be transported up from Earth. Paper is a dense material averaging between forty-four and seventy-two pounds per cubic foot (water has a density of about sixty-two pounds per cubic foot). A roll of toilet paper or a box of facial tissues thus becomes valuable, and you won't be able to use either as freely as you can here on Earth.

You may be permitted to take one or two favorite books with you, but may decide not to because you'll have the entire contents of many large libraries at your beck and call through the ubiquitous data base that's available at any habitat's computer terminals.

Because of this, you probably won't write letters and, even if you have an administrative or managerial job, you won't deal in paperwork. You'll instead use things that are cheap and readily available in any habitat: computers and communications.

Clothing will also be more valuable because the technologies of weaving and dyeing won't be adapted to space industry immediately. They're old technologies on Earth and, in their current and well-developed forms, require gravity. Fireproof or fire-resistant fabrics will be necessary in many habitats. Some types of cloth won't be permitted in certain habitats because of their flammability, high-temperature characteristics, or outgassing qualities. For example, some plastic-based or synthetic fabrics won't burn but instead melt and adhere to human skin, creating a human-safety problem during fire emergencies.

In a weightless habitat, shoes as you know and wear them on Earth become something else: protection against stubbing your toes. You may be issued cloth slippers or sandals with adhesive soles resembling Velcro or having a hard sole with a cleat that will slip into any of the foot restraints in the habitat.

If you like necklaces or neck chains, you'll have to learn to like something else, because any cordlike thing around your neck won't lie on your upper torso like a neck chain. It will float and may get tangled with some part of the habitat, becoming a noose.

No Smoking

Although there will be many different kinds of recreation available in all habitats as well as many ways to relax, let off steam, and get laid back, it's probable that the rules of most habitats won't permit cigarette smoking at any time, and not primarily for personal health reasons. Tobacco smoke contains some of the gooeyest brown gunk you could ever imagine. Many small-aircraft airplane pilots don't permit their passengers to smoke in the cabin because, in spite of filters, the gooey tars of tobacco smoke get into the delicate ball bearings of gyros and precipitate on switch contacts and control linkages, necessitating frequent and expensive replacements and cleaning. Marijuana smoking hasn't yet reached the level of tobacco use, so the physical effects of pot smoke on delicate equipment isn't as well known yet.

In fact, tobacco and other forms of smoking should probably be considered in the same league as using drugs such as cocaine, amphetamines, and other current "social" drugs. Their use affects or impairs your mental capabilities, your reaction times, your judgment, and even your physical condition. Even if not on duty or working, you are at all times in great risk of danger in any space habitat. There are any number of things that could happen to create emergencies in which it will be absolutely necessary for you to be physically and mentally at your very best.

Alcohol

The governing group or managers of a habitat may permit what they consider to be a reasonable social intake of alcoholic beverages for the purposes of relaxation and recreation. If this is the case, however, just as in aviation there will be strictly enforced rules relating to the elapsed time between consuming an alcoholic beverage and engaging in space working activities. In aviation, the rule is "eight hours from bottle to throttle," but airline transport pilots may not consume alcoholic beverages within twenty-four hours of a flight. It may be extremely difficult to prevent alcoholic beverages from being smuggled into habitats, and it may be even more difficult to keep people from moonshining, because it's easy to make a vacuum still . . . and there's a lot of vacuum out there.

The method of enforcement of habitat rules and regulations will also depend upon the sort of habitat you live in and how it's governed and operated. It's unlikely that habitats will have dedicated police or rule-enforcement people at first. Habitats will be too small and the value of each person working in them will be too great to permit the luxury of a police force. But it will come eventually as habitat size increases and as the number of people living together in a single habitat increases.

Space justice is going to be as developmental as true space law.

To date, space law has concerned itself primarily with such obtuse and impersonal problems as frequency assignments, orbital slot allotments, liability, and other legal matters related to what amounts to an extension of the law of the sea and the law of the air. However, much of this is inapplicable to the growing field of space justice. The biggest problem faced by space law—now as also in the immediate future—is that space lawyers have tried to anticipate technology. Therefore, they've established legal principles based upon what amounts to technological speculation, much of which turns out, at best, to be wrong, incomplete, and inapplicable to what really happens.

True space law will begin to develop once people are living in space, because the basis of all legal codes and actions is the nonviolent resolution of conflicts between people.

Initially, the legal systems used in various habitats will be straightforward extensions of whatever legal systems and organizations are utilized by the earthbound owners and managers of the habitat. "He who has the gold makes the rules." But, since space living involves new life-styles and the definite possibility of creating new and hitherto unsuspected interrelationships between people and between individuals and their institutions, the legal systems of various habitats will slowly develop, evolve, and mature.

Social Experiments in Space

Other social systems will also develop, evolve, and mature in space as time goes on.

Space will offer the social sciences one of the very best opportunities they've ever had to engage in long-term studies of human interactions, how institutions behave, and how both individuals and institutions react to known changes in the environment and to the situations in which they exist. Such studies are difficult here on Earth because of the continual interaction between people and institutions. No social "experiment" can really be isolated or even semi-isolated so that the investigators can get a solid grip on all the variables.

Although this is also going to be the case in the early years of space habitation because of the proximity of habitats to Earth, their dependence upon earthly supplies of oxygen, food, and other consumables, and the quick and easy communications with Earth, the evolution of space habitats and colonies farther from Earth and therefore far more independent and self-sufficient will provide the social scientists with growing opportunities to lend rigor to their fields by true social experimentation.

However, this sort of thing is in the future and perhaps something that will be of immediate interest to your children if they are born in and live in space. In the meantime, you're going to find out all the little tricks and gimmicks about living in space and interacting with your fellow human beings there. You'll be on a frontier that's more than the physical frontier of space; it's also the frontier of the social sciences.

16 Envoi—The Giant Leap for Mankind

When Neil A. Armstrong first stepped on the surface of the Moon at 2:56:20 Universal Time on July 21, 1969, his now-famous words echoed around the Planet Earth. He knew what he was saying, and he knew the deeper meaning of it. But it took more than a decade for people to really understand what Armstrong meant because he wasn't speaking for the history books as much as for the ages and generations to come.

You are doing much the same, in actions rather than in words. By going into space to live and work, you're taking part in something as vitally important to the future of the human race as anything that has happened in history to date.

People have been building civilization for at least 5,000 years and perhaps longer. We're still involved in the process of building institutions and civilization. You'll take this heritage with you to build a civilization in space. But the results of the civilization you and others like you build in space may be as different from what has been developed on Earth as those first primitive city states of Sumer and Assyria were from the village communities of India that preceded them.

Civilization is, literally, the art of living together in cities; that's what the roots of the word mean. Historian Will Durant defined civilization as "social order promoting cultural creation. Four elements constitute it: economic provision, political organization, moral traditions, and the pursuit of knowledge and the arts. It begins where chaos and insecurity end. For when fear is overcome, curiosity and constructiveness are free, and man passes by natural impulse toward the understanding and embellishment of life."

259

What you're doing in space is exactly that—providing for yourself and others in the economic area through your work and what you produce of value to others, organizing and operating the necessary institutions for political and cultural growth, and pursuing knowledge and the arts in scientific research or in technology because the full scope of the arts now includes these two elements.

You won't immediately step into a full-blown space civilization in which huge colonies spring up as if from nowhere. You'll have to create it, and it will take time. Civilization on Earth has taken at least fifty centuries to reach its present globe-spanning level. But you'll be able to create the space civilization faster because of the great storehouse of historic knowledge you possess as a foundation plus a communications system that operates at the speed of light.

The evidence of this storehouse of knowledge points toward the fact that we're in an era of great transition, perhaps the greatest age of change that has occurred in the last fifty centuries of recorded history. As the late Dr. Herman Kahn observed, "Two hundred years ago the human race almost everywhere was few, poor, and largely at the mercy of the forces of nature whereas two hundred years hence, barring some perverse combination of bad luck and bad management, the human race should be almost everywhere numerous, rich, and largely in control of the forces of nature." Space habitation may be an integral part of this transition.

To get some perspective on this, consider: It has been about two hundred years since humans began to fly in the air. Before that, people had left the surface of the Earth only when they jumped from a higher hill to a lower one. Before that, no one had ever ascended into the air like a bird. The first flight by François Pilâtre de Rozier in a hot-air balloon built by the brothers Jacques Etienne and Joseph Michel Montgolfier took place near Paris, France, on October 15, 1783. Benjamin Franklin, who was there as the United States Ambassador to France as well as a scientist in his own right, is reported to have told a querulous spectator of this event, "You ask of what use is this? Sir, of what possible use is a newborn babe?" Franklin, one of the most intelligent and educated men of his time, couldn't possibly have imagined what takes place two hundred years later every day at John F. Kennedy International Airport, where thousands of people embark on a short and comfortable journey across the Atlantic Ocean—a journey that takes hours in contrast to six weeks of hazardous sailing-ship passage.

This may sound a bit presumptuous, written as it is in the early dawn of the space civilization. But, recalling Franklin's words, I personally would rather err by being too presumptuous than to join thousands of other conservative and "right thinking" people who've turned out to be far too conservative in their thinking about the future as well as about what we're really going about doing.

Many people may consider your desire to go into space as an attempt to escape the "harsh realities" of living on Earth. But, as we've discussed, the "harsh realities" are not on Earth but in space, where no escape from them is possible without knowledge and technology. Living in space is no easy life and it won't be for decades to come.

Space as an environment for humanity is just as dangerous, hazardous, and deadly as most of the environments of Earth—the broad savannahs of central Africa, the rain forests of the tropics, the sprawling dry deserts, the high, cold

260

mountain passes, the seemingly endless expanses of the oceans, the lifeless tundra of the polar regions, and the envelope of air that surrounds our home planet. Space is just as different and dangerous to live in as the dry land was to those proto-amphibians who somehow managed to get stranded on the beach when the Silurian tide went out.

Even though you may at first go into space for several months at a time, you'll probably be living there for extended periods of time as our knowledge of space medicine and space living matures through experience. For this task, you're well equipped with a physical body and a large reasoning brain that have evolved and overcome all of the hazards of the terrestrial environment, making you a member of the most powerful species currently on Earth. You'll be able to use these faculties to do the same things in space if you know and understand what you are and where you came from.

You're also well equipped from the social point of view. Your ancestors learned to live together in the teeming cities of Europe, Asia, and America. You must now learn how to live with others in the far cleaner and more thoughtfully designed proto-cities of space. You'll be among those who build the space cities—which will look nothing like the cities of Earth and nothing like the way many forecasters now envision them. New institutions will come into existence. Eventually—it will take years—a self-sufficient culture will gradually emerge in space, one that's no longer tied to Earth for consumable supplies but that is self-supporting. But there will always be commercial traffic with Earth, an exchange of things of value, because that is why the space habitats were created in the first place. Therefore, although the space civilization you're building will eventually become less dependent upon Earth, it will be a very long time before the people of space don't need the people of Earth, and vice versa.

You'll not only be taking four million years of human development with you, but also the accumulated knowledge of the state of the art in every area of human endeavor. You'll operate in space in full accordance with your strong points as a human being as well as with your shortcomings, both individual and collective.

Most important for the long-term health and success of the human race, you'll be one of those who began the long and enormously important task of taking humanity's eggs out of one fragile planetary basket. The first step in this task will be achieved when enough people with accumulated critical know-how exist in space to survive a worldwide thermonuclear holocaust.

If, as a result of Kahn's "perverse combination of bad luck and bad management," that or a similar earthly catastrophe occurs, not only will the human seed have survived in space, but you and your progeny will be able to return like gods from the skies to help rebuild the world.

The next step in this task involves the universal expansion of humanity to the point where the human race could survive even if the sun itself dies. But at this time the subject of space habitation beyond our own solar system is much too speculative, in scientific and technical terms, for a book such as this.

You're going into space for personal reasons of your own, but there are also deeper reasons why you and all the rest of us are doing this new thing. The personal reasons are important to you; the impersonal reasons are important to the rest of the human race as it grows into maturity as a species.

INDEX

263

268

270

271